EPA420-R-01-001
January 2001

EPA Guidance:

Improving Air Quality Through Land Use Activities

Transportation and Regional Programs Division
Office of Transportation and Air Quality
U.S. Environmental Protection Agency

TABLE OF CONTENTS

Chapter 1   **Introduction** .... -1-
       1.1   Why did EPA develop this guidance? .... -1-
       1.2   What does this guidance do? .... -1-
       1.3   Who will use this guidance? .... -2-
       1.4   How do I know if this guidance is applicable to my community? .... -3-
       1.5   What are the key concepts discussed in this guidance? .... -4-

Chapter 2   **Land Use Planning** .... -5-
       2.1   What is land use planning? .... -5-
       2.2   Who gets involved in land use decision making? .... -5-
       2.3   What are the some of the tools used in land use planning? .... -8-

Chapter 3   **Air Quality and Transportation Planning** .... -10-
       3.1   What is the concern about air quality and transportation in the United States?-10-
       3.2   Has air pollution from transportation sources improved? .... -11-
       3.3   What is the air quality planning process? .... -13-
       3.4   Who gets involved in the air quality planning process? .... -14-
       3.5   What is the transportation planning process? .... -14-
       3.6   Who is involved in the transportation planning process? .... -15-
       3.7   How are the air quality and transportation planning processes linked? .... -15-

Chapter 4   **Linking Land Use, Air Quality, and Transportation Planning** ...... -16-
       4.1   What is the relationship between land use, transportation,
             and air quality? .... -16-
       4.2   In what ways does urban form impact travel activity? .... -16-
       4.3   How can urban form be changed to improve air quality? .... -20-
       4.4   What kinds of actions are needed to make "smart growth"
             strategies achievable? .... -22-

Chapter 5   **Accounting for emission reductions from land use activities** .... -27-
       5.1   What kinds of land use activities can be accounted for in SIPs and
             conformity determinations? .... -27-
       5.2   How are land use activities incorporated into the air quality and
             transportation processes? .... -27-
             5.2.1   What are land use planning assumptions? .... -28-
             5.2.2   What is the travel demand forecasting process? .... -29-
             5.2.3   What is the emissions modeling process? .... -30-
       5.3   How are land use activities incorporated into the SIP and the conformity
             determination? .... -31-
       5.4   What are the ways that I can account for land use activities? .... -31-

Chapter 6   **Including land use activities in the initial forecast of future emissions**
               in the SIP .... -33-
       6.1   What is the initial forecast of future emissions? .... -33-
       6.2   When is an initial forecast of future emissions made? .... -33-

	6.3	How can I account for "smart growth" activities in the land use assumptions that are made for the SIP?	-33-
	6.5	What is "double counting?"	-37-
	6.6	What else should I consider when including land use activities in my initial forecast of future emissions?	-38-

Chapter 7   **Including a land use activity as a control strategy in the SIP** .... **-39-**

	7.1	What is a control strategy?	-39-
	7.2	When would I include land use activities as control strategies in the SIP?	-39-
	7.3	How can I account for land use activities as control strategies in my SIP?	-39-
	7.4	Land use activities as Traditional Control Strategies	-39-
		7.4.1 What are Traditional Control Strategies?	-40-
		7.4.2 What are the existing statutory requirements for including a land use activity as a traditional control strategy in a SIP?	-40-
		7.4.3 What happens if I have included a land use activity as a traditional control strategy in a SIP, and now I have information that the land use activity is not occurring?	-43-
		7.4.4 What else should I consider when including a land use activity as a traditional control strategy in a SIP?	-43-
	7.5	Land use activities and the Voluntary Mobile Source Emission Reduction Program policy (VMEP policy)	-44-
		7.5.1 What is the Voluntary Mobile Source Emission Reduction Program (VMEP) policy?	-44-
		7.5.2 What are the existing statutory requirements for including a land use activity as a VMEP control strategy in a SIP?	-44-
		7.5.3 When should I use the VMEP policy to include a land use activity in a SIP?	-45-
		7.5.4 What happens if I have included a land use activity as a VMEP control strategy in the SIP, and I now have information that it is not occurring?	-46-
		7.5.5 What else should I consider when including a land use activity as a VMEP control strategy in a SIP?	-46-
	7.6	Land Use Activities and the Economic Incentive Programs policy (EIP policy)	-47-
		7.6.1 What is the Economic Incentive Programs (EIP) policy?	-47-
		7.6.2 How is the EIP policy related to the VMEP policy?	-48-
		7.6.3 When should I use the EIP Policy to include a land use activity in a SIP?	-49-
	7.7	What steps are necessary for quantifying land use activities as traditional, VMEP, or EIP control strategies?	-49-
	7.8	Transportation Control Measures as control strategies	-52-
		7.8.1 What are Transportation Control Measures?	-52-
		7.8.2 How are Transportation Control Measures and land use activities related?	-52-

Chapter 8   **Including land use policies or projects in the conformity determination without having them in a SIP** .... **-55-**

	8.1	What is a conformity determination?	-55-
	8.2	How is conformity demonstrated?	-55-

8.3	Does this guidance impose new requirements for including land use activities in a conformity determination?	-56-
8.4	If I have included a land use activity in a SIP, does it have to be included in the conformity determination?	-56-
8.5	Can I account for the emissions benefits of land use activities in a conformity determination without having them in a SIP?	-56-
8.6	How are land use activities included in the conformity determination?	-56-
8.7	What are the transportation conformity rule's requirements for land use assumptions?	-57-
8.8	How are the land use assumptions in a conformity determination reviewed?	-59-
8.9	What are control strategies?	-59-
8.10	What are the conformity rule's requirements for control strategies?	-60-
8.11	How do I determine whether a land use activity is a land use assumption or a control strategy?	-60-
8.12	What are some examples of land use activities that fit in each category?	-61-
8.13	What is "double counting?"	-62-
8.14	What if a land use activity is too small to have an impact on the outcome of travel demand modeling?	-63-
8.15	What if our area doesn't use a travel demand model for transportation planning?	-63-
8.16	What are the advantages of accounting for land use activities in the conformity determination without having them in the SIP?	-63-

Chapter 9 **Additional considerations when accounting for land use activities in the SIP or the conformity process** .... **-65-**

9.1	How can I determine whether or not my land use activities might have air quality benefits?	-65-
9.2	How will the time frame for implementing the land use activities affect which accounting option I choose?	-66-
9.3	What other important issues should I be aware of in quantifying air quality benefits?	-67-
	9.3.1 Accounting for interactions between land use activities	-67-
	9.3.2 Quantifying land use activities individually or as a group	-67-
	9.3.3 Using conservative estimates	-68-
	9.3.4 Taking into account the scale of the land use activity	-68-
9.4	How will EPA assist me with quantification?	-68-

Appendix A	**Examples of Land Use Policies and Strategies**	**A-1**
Appendix B	**Related Internet Web Sites**	**B-1**
Appendix C	**Related Work Efforts**	**C-1**
Appendix D	**Glossary of Terms**	**C-1**
Appendix E	**List of Acronyms**	**D-1**

Appendix F References to Relevant Policies, Guidance Documents, and General
Information Sources E-1

Appendix G Regional and State Contacts F-1

EXECUTIVE SUMMARY

Recently states and local communities have passed hundreds of ballot initiatives preserving open space, increasing development around transit, and providing for increased brownfield redevelopment. Each of these places has had different reasons--economic, environmental or community goals--for pursuing a chosen development path. Environmentally, these decisions can help communities reduce vehicular emissions, improve water quality, and remediate contaminated lands.

States and communities are interested in accounting for the air quality benefits of their development choices. This guidance presents the conditions under which the benefits of land use activities could be included in air quality and transportation planning processes. The United States Environmental Protection Agency (EPA) intends that this guidance be an additional tool to encourage the development of land use policies and projects which improve livability in general, and air quality in particular. This effort is intended to complement the efforts of states and local areas, and to provide guidance, flexibility and technical assistance to areas that wish to implement these measures and use them towards meeting their air quality goals.

This guidance document is a non-regulatory interpretation and clarification of EPA's policies and practices relating to treatment of land use activities and is consistent with Section 131 of the Clean Air Act, which states, "Nothing in this Act constitutes an infringement on the existing authority of counties and cities to plan or control land use, and nothing in this Act provides or transfers authority over such land use."

The goal of this guidance is to assist air quality and transportation planners in accounting for the air quality impacts of land use policies and projects which state and local governments *voluntarily* adopt. EPA is providing this guidance to give flexibility to state and local governments by expanding the number of strategies an area can use to meet its air quality planning requirements. Properly modeled and quantified land use activities have the potential to help local areas meet their air quality goals, and impact the quality of life of all citizens. Guidance on quantifying land use strategies not discussed in this document may be addressed in future documents.

In general, states can account for the air quality benefits of land use activities for nonattainment and maintenance areas in one of three ways:

- ♦ Including land use activities in the initial forecast of future emissions in the SIP;

- ♦ Including land use activities as control strategies in the SIP; and

- ♦ Including land use activities in a conformity determination. without including them in the SIP.

The guidance consists of three sections:

Section 1 of the document provides an overview of the land use, air quality, and transportation planning processes, and discusses the links between these processes. This sections is designed primarily for those who do not have much experience with these processes.

- Chapters 2 and 3 provide background information on the air quality and transportation planning processes and the land use planning process.

- Chapter 4 conveys the connection between each of these three processes, and the ways that governments, developers, and citizens can affect these processes to potentially improve quality of life in general, and, more specifically, transportation options and air quality.

Section 2 of the document describes the appropriate application of existing EPA policies when accounting for land use activities in State Implementation Plans and conformity determinations. This section also discusses general quantification guidelines for land use activities and is designed primarily for air quality and transportation planning professionals.

- Chapter 5 discusses how land use is incorporated into the travel demand modeling process.

- Chapters 6, 7, and 8 discuss three key ways that the beneficial air quality impacts of some land use and transportation decisions can be accounted for in the transportation and air quality planning processes.

- Chapter 9 discusses special considerations when accounting for land use activities in these planning processes

Section 3 of the document contains appendices containing further information and resources for users of this guidance.

Additional guidance materials will be released over time on specific topics, such as methodologies for quantifying the impacts of specific land use policies and projects like infill development, brownfields redevelopment, and transit oriented development.

For further information, please contact:

> Office of Transportation and Air Quality
> TRAQ Center Information Request Line
> 734-214-4100
> http://www.epa.gov/oms/transp/traqsusd.htm

SECTION 1: OVERVIEW OF THE LINK BETWEEN LAND USE, TRANSPORTATION, AND AIR QUALITY

Section 1 is designed to provide a primer for readers who may not be familiar with the concepts and processes that are related to accounting for air quality impacts of land use activities. Readers who are more familiar with these topics may wish to skip to Section 2, which covers specific policy and technical considerations.

Chapter 1	**Introduction**	-1-
1.1	Why did EPA develop this guidance?	-1-
1.2	What does this guidance do?	-1-
1.3	Who will use this guidance?	-2-
1.4	How do I know if this guidance is applicable to my community?	-3-
1.5	What are the key concepts discussed in this guidance?	-4-
Chapter 2	**Land Use Planning**	-5-
2.1	What is land use planning?	-5-
2.2	Who gets involved in land use decision making?	-5-
2.3	What are the some of the tools used in land use planning?	-7-
Chapter 3	**Air Quality and Transportation Planning**	-9-
3.1	What is the concern about air quality and transportation in the United States?	-9-
3.2	Has air pollution from transportation sources improved?	-10-
3.3	What is the air quality planning process?	-12-
3.4	Who gets involved in the air quality planning process?	-13-
3.5	What is the transportation planning process?	-13-
3.6	Who is involved in the transportation planning process?	-14-
3.7	How are the air quality and transportation planning processes linked?	-14-
Chapter 4	**Linking Land Use, Air Quality, and Transportation Planning**	-15-
4.1	What is the relationship between land use, transportation, and air quality?	-15-
4.2	In what ways does urban form impact travel activity?	-15-
4.3	How can urban form be changed to improve air quality?	-19-
4.4	What kinds of actions are needed to make "smart growth" strategies achievable?	-21-

CHAPTER 1 INTRODUCTION

1.1 WHY DID EPA DEVELOP THIS GUIDANCE?

In recent years, many of EPA's stakeholders have explored using land use activities as strategies for improving air quality. These stakeholders, including state and local planning agencies, have suggested that EPA provide guidance on how to recognize land use strategies in the air quality planning process that result in improvements in local and regional air quality. In a survey conducted by EPA in 1998[1], staff and managers in state air agencies and regional transportation planning agencies said that being able to quantify and account for the air quality impacts of beneficial land use activities would:

- Encourage funding for research into the impacts of such activities,

- Educate state and local government officials about land use planning as a tool for achieving clean air, and

- Add support to these kinds of activities in regional and local debates.

This guidance document is designed to describe how you can use existing EPA regulations and policies to account for the air quality benefits of land use activities that encourage travel patterns and choices that reduce vehicle miles of travel, and consequently reduce emissions from motor vehicles in your communities. This document lays out general guidance on quantifying the potential benefits of land use activities that your area may choose to adopt. EPA will provide additional guidance on quantifying benefits from specific types of land use strategies in the future.

1.2 WHAT DOES THIS GUIDANCE DO?

The goals of this guidance are to:

- **Describe the options** for accounting for the air quality benefits of land use activities in the air quality planning and transportation planning processes (i.e., state implementation plans (SIPs), and conformity determinations),

- **Help you determine which option is appropriate** for a chosen land use activity, and

- **Help you model the air quality impacts** of land use activity.

---

[1] USEPA, *Background Information for Land Use SIP Policy*, EPA420-R-98-012, October 1998.

This guidance is *not* a regulatory document. This guidance is consistent with Section 131 of the Clean Air Act, which states:

> "Nothing in this Act constitutes an infringement on the existing authority of counties and cities to plan or control land use, and nothing in this Act provides or transfers authority over such land use."

This guidance will help air quality and transportation planners account for emission reduction impacts of beneficial land use activities which state and local governments *voluntarily* adopt. It is intended to inform state and local governments that land use activities which can be shown (through appropriate modeling and quantification) to have beneficial impacts on air quality, may help them meet their air quality goals. EPA is providing this guidance to give flexibility to state and local governments by expanding the number of strategies an area can use to meet its air quality planning requirements.

Another purpose of this document is to **emphasize the importance of strong coordination and cooperation** among the many parties involved in land use, transportation, and air quality planning. Communities in all parts of the country are grappling with deciding what kinds of growth patterns best meet their goals. In many of these areas, state air agencies are struggling to identify new ways to meet their state's air quality goals, and state transportation planners are attempting to meet community transportation needs with projects that are consistent with existing air quality goals. Land development patterns influence travel decisions, and therefore have a direct impact on air quality. Therefore, collaboration and early involvement of the local, regional, and state government agencies, as well as members of the public, environmental and community action organizations, and the development community, are vital to ensuring that the wide array of community goals are adequately considered.

1.3 WHO WILL USE THIS GUIDANCE?

Both the air quality and transportation planning processes involve consideration and estimation of the effects of land use on travel activity. While local governments have the primary responsibility for decisions regarding land use, there are three main government agencies that are responsible for assessing regional air quality and the effects of transportation on air quality:

> **State and Local Air Quality Agencies** – agencies at the state or local level that prepare air quality plans (known as State Implementation Plans, or SIPs) will use this guidance for quantifying the air quality benefits of land use activities in SIPs;
>
> **Regional Transportation Planning Agencies** – these agencies, which most often are federally-designated Metropolitan Planning Organizations (MPOs) and Councils of Government (COGs), prepare long-term transportation plans and shorter-term transportation improvement programs for metropolitan areas and are responsible for showing that transportation activities "conform" to the goals of air quality plans. They will use this guidance for quantifying the air quality benefits of land use activities in conformity determinations; and
>
> **State Departments of Transportation** – these agencies work with regional transportation planning and air quality agencies to develop and evaluate transportation plans and programs, and are usually responsible for demonstrating that transportation activities in rural areas conform to the goals of state-wide air quality plans.

This guidance will be used primarily by those agencies responsible for air quality and transportation planning. However, other parties that may be interested in the ideas, policies and technical issues presented in this guidance include:

- Local government agencies
- Regional agencies
- State agencies
- Federal agencies
- Private developers
- Academia
- Citizens and community organizations
- Financial Institutions

1.4 HOW DO I KNOW IF THIS GUIDANCE IS APPLICABLE TO MY COMMUNITY?

This guidance is most relevant for areas that are designated nonattainment or maintenance areas for the air pollutants ozone, PM-10, CO, and/or NO_2.

> **Nonattainment area:** a geographic region of the United States that the EPA has designated as not meeting the National Ambient Air Quality Standards (NAAQS) for specific air pollutants.
>
> **Maintenance areas:** an area previously designated nonattainment, which has since met the national standards and has an EPA approved maintenance plan covering at least 10 years.

Under the Clean Air Act, states are required to submit plans to EPA showing how they will meet their air quality goals. They must also show that any planned transportation activities are in harmony with those goals.

In addition, with the recent release of the EPA Regional Haze rule, states may want to consider the impacts of land use on transportation, as vehicle emissions are contributors to regional haze pollution.

While this guidance is designed primarily to help areas achieve their air quality goals to meet federal standards, local, regional, and state government staff in areas that are not designated as nonattainment areas may also find this document useful as they explore ways to secure their attainment status in the future and determine potential CO_2 impacts of their land use choices.

1.5 WHAT ARE THE KEY CONCEPTS DISCUSSED IN THIS GUIDANCE?

In this guidance, we discuss the links between the land use, transportation, and air quality planning processes. State and local governments use these processes to make choices about community growth, economic development, transportation infrastructure support and development, and protection of environmental quality. Several concepts are discussed throughout the document; brief definitions are provided below. These concepts are discussed in greater detail later in this document

Air quality planing
- The process by which state, and in some cases, regional, air quality planning agencies assess current and future air quality conditions and determine the "control strategies"needed to reduce emissions and improve air quality. These agencies prepare State Implementation Plans (SIPs) and submit them to EPA for approval.

Transportation planning
- The process by which state and local transportation agencies, along with metropolitan planning organizations, assess needs for future transportation infrastructure such as roads and transit systems. Federal regulations require states to demonstrate that planned transportation activities are consistent with or "conform" to the air goals outlined in the SIP.

Land use planning
- The process by which local governments plan for future growth in communities and decide where and how development should occur within local boundaries. In some cases, regional planning agencies work with local governments to coordinate the planning efforts of neighboring municipalities.

Land use activities
- Land use activities include all of the various actions that state and local governments or other entities take which affect the development of land use in a community or region. These land use activities result in patterns of land use that influence the transportation choices people make. In this guidance, *land use activities* that reduce reliance on motor vehicles (e.g., through shortening trip lengths or increasing accessibility of alternative modes of transportation) and that can also be shown to have air quality benefits may be accounted for in the air quality and transportation planning processes.

    Land use activities include *land use policies*, defined as specific policies, programs, or regulations adopted or operated by government agencies and *land use projects* defined as specific developments.

Accounting for land use in the air quality and transportation processes
- Where planning agencies can demonstrate ,through modeling, that land use activities can be reasonably expected to have a positive impact on air quality, they can account for those benefits in the air quality and/or the transportation planning process.

CHAPTER 2 LAND USE PLANNING

2.1 WHAT IS LAND USE PLANNING?

Land use planning is a process through which government agencies assess current physical, social, and economic conditions within an area, project future trends based on this information, and consider strategies and alternatives for developing land that meet the needs of the community at large. Part of the planning process involves making decisions about the physical characteristics and layout of development in a metropolitan area, including development of transportation infrastructure, and building design and orientation.

Land use planning sometimes involves consideration of a multitude of community concerns, including environmental concerns such as air quality.

2.2 WHO GETS INVOLVED IN LAND USE DECISION MAKING?

Local, regional, and state government agencies all have a role in land use decision-making. In addition, individuals, community organizations, and developers play important roles in the process. The roles of these parties in the land use planning process are discussed in the following sections.

Role of local governments

In most states, land use decisions are primarily made at the local level. Metropolitan areas are geographic regions that are comprised of a central city of at least 50,000 people, or a U.S. Census Bureau-defined urbanized area and a total population of at least 100,000 (75,000 in New England), including suburbs surrounding the central city, edge cities, and rural fringe areas.[2] These metropolitan areas are comprised of a number of local government bodies, such as counties, municipalities, cities, and townships, as well as special service districts and school districts.

Local governments are generally responsible for, and have authority over, various land use decisions within their borders. They have the authority to issue permits for development and control where and how development occurs. Since each of these bodies may govern some aspects of land use decisions, there always exists the possibility that the goals of one body may conflict with those of another.

---

[2] This definition was provided by the U.S. Bureau of the Census. For more information, visit their web address at http://www.census.gov/population/www/estimates/aboutmetro.html.

> **LOCAL GOVERNMENT BODIES THAT INFLUENCE LAND USE**
>
> There are several different levels of local government that each have responsibility for making separate decisions about land use.
>
> ♦ **Municipalities**: individual villages, towns, cities and boroughs; usually have control over land use planning and zoning within their boundaries.
>
> ♦ **Townships**: counties can be comprised of townships, which in some cases have control over land use planning, zoning, and roads.
>
> ♦ **Counties:** often governed by boards called planning commissions, which may have the responsibility for controlling land use planning, zoning, and expansion of city limits; often are also responsible for road and bridge construction and maintenance. May have decision making role, or may serve as advisory bodies for municipal governments.
>
> ♦ **Special districts**: some services, such as water and sewage treatment, are governed by districts and authorities that do not necessarily correspond with municipal boundaries. Decisions regarding these services are often beyond the control of the municipalities.
>
> ♦ **School districts**: the boundaries and governance of school districts are often different from municipal boundaries.

Role of regional government

There are typically two types of regional planning agencies that are connected to land use planning issues. These agencies are known as councils of governments *(COGs)* and metropolitan planning organizations (MPOs); in some areas, the MPO and the COG are the same agency.

Councils of Governments (COGs) help local governments and the state by preparing land use forecasts, with input from the local governments, which are used in transportation and air quality planning, as well as economic planning. COGs serve advisory roles, and their decisions are not legally binding.

Metropolitan Planning Organizations (MPOs), together with the state, are responsible for conducting the continuing, cooperative, and comprehensive regional transportation planning process under the Federal Highway Administration's planning rules[3]. MPOs generally are responsible for distribution of federal transportation funding. In its planning activities, the MPO must consider the relationship between planned transportation activities and air quality goals in both the near term and in the long term.

These regional planning agencies play important roles by providing land use and transportation data and analyses of the regional impacts of alternative land use scenarios, but generally have limited influence on land use decisions. Two notable exceptions can be found in Oregon and Georgia. The State of Oregon has created a voter-elected regional government body known as Metro, which has some legally binding planning authority over land use decisions in the Portland metropolitan area. In 1999, the governor of Georgia created a planning body known as the Georgia Regional Transportation Authority (GRTA), which has the authority to approve local land use/transportation plans, and can influence the transportation planning process by withholding state funding.

---

[3] Federal Highway Administration, Federal Transit Administration, 23 U.S.C. 134 and 49 U.S.C. 5303.

Role of state government

States generally delegate land use decision making authority to local governments. While states usually have little direct impact on those decisions, numerous state programs governing taxes, infrastructure funding, highways, and community investment indirectly exert a strong influence on land use decisions. Many states are beginning to evaluate state-level policies that pull development away from community centers and are looking at ways to transform those policies into incentives for compact, community-based development.

Role of federal government

Although federal agencies are not involved in local land use decisions, federal statutes such as environmental laws, tax codes, federal mortgage lending policies, and transportation infrastructure policies can influence local land use planning.

Examples of such policies include assessment requirements in the National Environmental Policy Act (NEPA), transportation planning requirements found in U.S. Department of Transportation regulations, and specifications on property use included in the EPA's Superfund regulations.

Role of developers

Private developers can propose a variety of development strategies and projects to local planning bodies, which can approve or disapprove the proposals. Developers can partner with communities to design developments that are in keeping with local economic, social, and environmental goals.

Role of citizens

Individuals and community organizations can significantly impact the direction of policy development through community organizing, ballot initiatives, creation and support of alternative development plans and projects, and participation in public hearings and comment periods related to specific development proposals.

2.3 WHAT ARE THE SOME OF THE TOOLS USED IN LAND USE PLANNING?

Local-level land use planning usually involves the development of a guiding document referred to by many names, including *comprehensive plan*, *master plan*, or *general plan*. This document is used to develop a "blueprint" for future development. In some states, local governments are required to create a comprehensive plan, while in others, such planning is optional. The comprehensive plan can be a legally binding document for local decision making, or it can be an advisory tool.

Beyond the comprehensive plan, local governments also develop specific land use regulations that are legally binding. These include *zoning ordinances* and *subdivision regulations*. Zoning ordinances are written specifications as to how and where development can occur. Characteristics such as height and size of buildings, building setbacks, lot and yard size, provisions for sidewalks and street widths, and minimum or maximum parking allowances are governed by zoning ordinances. Subdivision regulations govern the design principles involved in development of new subdivisions and define how parcels of land will be developed.

Local governments can influence site design by employing *development standards* through a design review process. In this process, new projects are brought before a design review board for approval; this review may include addressing air quality issues.

Local governments may also develop *incentive programs*. Incentives may be monetary or non-monetary. For example, local governments may offer rewards to developers who build in desired locations or who include certain design features in new projects. Local governments may also establish assessment districts where developers are required to pay for the cost of needed infrastructure improvements.

Regional and state government agencies can also influence development patterns. These agencies can develop region-wide comprehensive planning processes to help coordinate planning efforts among local governments. Regional and state government agencies can also create policies and programs to educate and encourage local governments to achieve desirable planning objectives, and can develop incentive programs that attract new development to desired areas. In some areas, regional and state agencies play a larger role in the land use planning process, and may possess greater authority over land use decision making.

While the federal government does not have jurisdiction over local land use decision making, federal statutes and funding policies do influence local land use decisions. Grant programs that assist states in redeveloping abandoned brownfields, earmarking federal funding assistance for "empowerment zones" in older urban areas, and partnerships between federal agencies and state and local governments to test land use planning tools are some examples.

Chapter 3 Air Quality and Transportation Planning

3.1 What is the concern about air quality and transportation in the United States?

Air pollution causes harm to humans, animals, plant life, water quality, property, and visibility. There are many different sources of air pollution, including naturally occurring sources (such as windblown dust, and volcanic eruptions), and man-made sources, such as stationary sources (e.g., factories, power plants) and transportation sources (e.g., cars, buses, planes, trucks, and trains).

The 1970 Clean Air Act (CAA)[4], and the subsequent 1977 and 1990 amendments, charged EPA with the task of establishing air quality standards based on maximum acceptable atmospheric concentrations of six air contaminants, known as *criteria pollutants*. All states must develop plans that demonstrate how they will attain and maintain the standards. In their plans, states must address emissions from motor vehicles.

Transportation sources are significant contributors to emissions of volatile organic compounds (VOCs) and oxides of nitrogen (NOx), the two major air pollutants related to smog. Transportation sources also contribute to the other criteria pollutants that EPA regulates. Table 1 shows the percentage of the total emissions inventory for each pollutant that is due to on-road and non-road vehicles and engines.

Table 1. Transportation Source Contribution to 1998 National Emissions Inventories of Criteria Pollutants[5]	
Criteria/ Precursor Pollutant	**% of total inventory (due to On-road and Non-Road Vehicles and Engines)**
Carbon Monoxide (CO)	79
Oxides of Nitrogen (NOx)*	53
Volatile Organic Compounds (VOCs)*	43
Particulate Matter (PM_{10})	19
Lead (Pb)	13
Sulfur Dioxide (SO_2)	7

*Note: NOx and VOCs are not criteria pollutants, but are precursor to the criteria pollutant ozone (O_3).

Transportation sources are also significant contributors to emissions of greenhouse gases, which are linked to global climate change. Transportation sources accounted for approximately 31% of total U.S. emissions of CO_2, a greenhouse gas, in 1998.[6]

---

[4] The Clean Air Act, 42 U.S. C. 7401 et seq.

[5] USEPA, 2000. *National Air Pollutant Emissions Trends 1900-1998*. EPA-454/R-00-002. USEPA Office of Air Quality Planning and Standards, Research Triangle Park, NC.

[6] USEPA 2000. *Inventory of Greenhouse Gas Emissions and Sinks: 1990-1998*. EPA 236-R-00-001 USEPA Office of Atmospheric Programs, Washington, DC.

3.2 HAS AIR POLLUTION FROM TRANSPORTATION SOURCES IMPROVED?

In general, yes. Much has been done to reduce emissions of air pollutants from cars and trucks over the last 25 years. These efforts have focused on the use of technology and tailpipe controls, and have been quite successful at reducing the emissions of criteria air pollutants from transportation sources. When compared to passenger cars in 1970, a vehicle today emits 60 to 90 percent less air pollution over its lifetime due to EPA's vehicle emissions standards.[7]

However, although EPA's emissions standards and the improved emissions control technology used to meet them are expected to continue to positively affect air quality, the number of miles being driven are expected to continue to increase.

FIGURE 1. TRENDS IN VEHICLE EMISSIONS AND VEHICLE MILES TRAVELED[8]

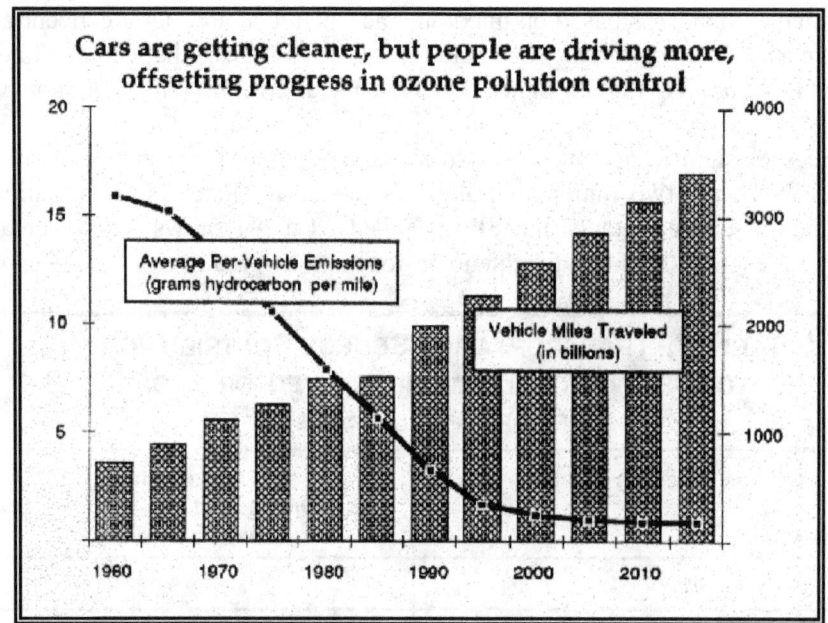

This increase in vehicle miles traveled (VMT) can be attributed to a number of changes in travel behavior. Since 1952, the number of cars and trucks in the US has more than quadrupled while the US population has less than doubled.[9] The average total annual mileage driven by Americans in 1995 (9,567 miles per year) was almost twice as high as it was in 1970 (4,587 miles per year).[10]

---

[7] US EPA Office of Air and Radiation, *Resource Information, Office of Transportation and Air Quality.* EPA420-F-00-004, February 2000.

[8] US EPA Office of Air and Radiation. Fact Sheet OMS-4 EPA 400-F-92-006, January 1993.

[9] American Automobile Manufacturers Association, *World Motor Vehicles Data* 1996 Edition, 1998.

[10] Oak Ridge National Laboratory, *Transportation Energy Data Book*: Edition 19, September 1999.

Furthermore, as Table 2 shows, people are taking more local trips and more long distance trips, and the trips are, on average, longer than they were 20 years ago.

TABLE 2. POPULATION AND PASSENGER TRAVEL CHARACTERISTICS IN THE UNITED STATES, 1977-1995[11]			
	1977	1995	Percentage Increase 1977-1995
Population (thousands)	219,760	262,761	20
Local Trips			
Annual local person trips (travel day) (millions)	211,778	378,930	79
Local person trips per capita, one way (per day)	2.4	4.3	47
Local person miles (millions)	1,879,215	3,411,122	82
Local person miles per capita (annually)	9,470	14,115	49
Local average trip length (miles)	8.9	9.0	1
Long Distance Trips			
Annual long distance person trips (millions)	521	1,001	92
Long distance trips per capita, round trip (per year)	2.5	3.9	56
Long distance person miles (millions)	382,466	826,804	116
Long distance miles per capita (annually)	1,796	3,129	74
Long distance average trip length (miles)	733	826	13

The majority of this travel activity occurs in personal automobiles. In many communities, alternative modes of transportation, such as transit and the infrastructure needed to make biking and walking safe and convenient, may be scarcely available, thus limiting personal transportation choices. Also, given average trip lengths of nearly 9 miles, walking and biking options are, in many cases, not viable.

---

[11]U.S. Department of Transportation. 2000. *The Changing Face of Transportation*, Table 5.1 Page 5-2.

FIGURE 2. TRANSPORTATION CHOICE BY MODE [12]

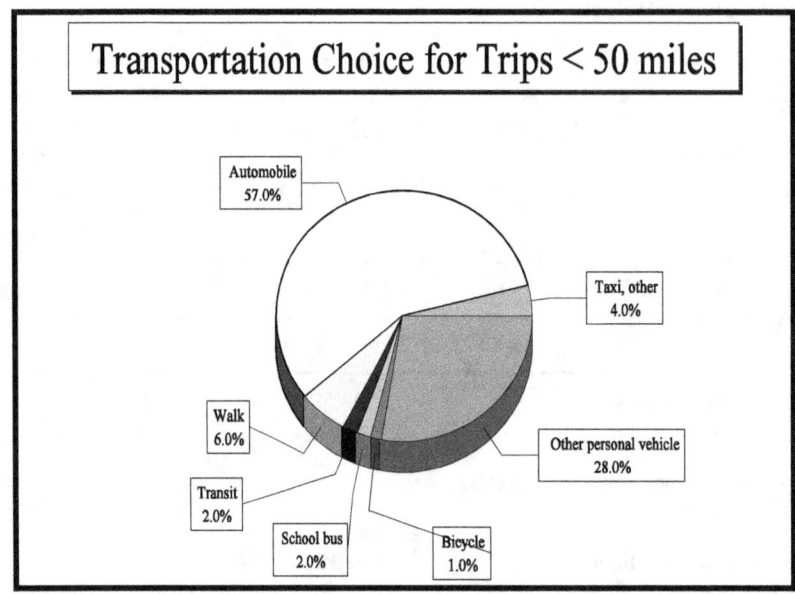

For the many areas of the country that are working hard to improve air quality, these trends toward more cars on the road, people driving more often, and increased trip lengths may offset the benefits gained from cleaner vehicles. To combat these trends, state and local government agencies seeking to reduce emissions from cars are increasingly looking not just at technological solutions, but at strategies to reduce driving. One such strategy is to explore ways to develop land within communities that will increase transportation options and provide people with more choices to use of alternative modes of travel and drive less.

3.3 WHAT IS THE AIR QUALITY PLANNING PROCESS?

The Clean Air Act directs state air quality agencies to prepare air quality plans (called **State Implementation Plans, or SIPs**). SIPs include estimates of future air quality and describes in detail a state's plans to achieve air quality goals. The approved SIP serves as the state's commitment to actions that will reduce air quality problems. A general description of the process follows.

A state's air quality agency (generally a branch of the state's environmental protection, natural resources, or public health department) prepares SIPs with input from the metropolitan planning organizations, industrial pollution sources operating in the state, and members of the public.

States submit SIPs for their nonattainment and maintenance areas, which can include both large metropolitan cities and/or rural areas. There are several kinds of SIPs that are required under different circumstances. The ones that are relevant to this guidance are control strategy SIPs (15% plans, rate-of-progress plans, and attainment plans) and maintenance plans. The control strategy SIPs must include an initial forecast of the area's future emissions -- that is, emissions in the year(s) addressed by the SIP that will result if no additional control strategies are implemented other than what is required by law. The SIP also must include a description of specific programs (or "control strategies") that will be used to achieve the needed emission reductions in the area, and calculate future emissions that result when the control strategies are considered. The area must adopt enough control strategies to show that it will:

---

[12] U.S. Department of Transportation (USDOT), Federal Highway Administration (FHWA), 2000. *Summary of Travel Trends: 1995 Nationwide Personal Transportation Survey.* Washington, DC.

- meet the standards in the future attainment year (known as "attainment demonstration" SIPs);

-or-

- continue to meet the standard after attainment is reached (known as "maintenance plan" SIPs).

Ozone areas have additional SIP requirements. Before attainment is reached, an ozone nonattainment area must:

- make progress in meeting the ozone standards (shown in SIPs called "15% plans," and "rate-of-progress plans").

The total allowable emissions in a control strategy SIP is the amount of emissions the area can have and still achieve the air quality goals of the SIP (progress, attainment, or maintenance of the National Ambient Air Quality Standards). The portion of the total allowable emissions that is allocated to highway and transit vehicle use is called the **motor vehicle emissions budget**, referred to hereafter as "the budget." The SIPs described above -- 15% plans, rate-of-progress plans, attainment demonstrations, and maintenance plans -- almost always contain these budgets. Motor vehicle emissions must stay within these budgets.

3.4 WHO GETS INVOLVED IN THE AIR QUALITY PLANNING PROCESS?

States, and in some cases, local air quality planning agencies, are responsible for creating and monitoring progress of the State Implementation Plan. State and local air quality planning agencies, and other organizations who are responsible for developing, submitting, or implementing provisions of a SIP consult with each other, with other state and local agencies, and with the appropriate local or regional offices of EPA and DOT. Air quality agencies and transportation planning agencies work together to model and assess the transportation network's impact on emissions from automobiles. In addition, citizens are allowed to review and comment on the proposed plan.

3.5 WHAT IS THE TRANSPORTATION PLANNING PROCESS?

The U.S. Department of Transportation planning regulations require areas with more than 50,000 residents to have a metropolitan planning organization (MPO). MPOs are responsible for transportation planning, specifically the creation of the area's long-range *transportation plan* and its shorter-term *transportation improvement program* (TIP).

The transportation plan is a long-term plan for maintenance and improvement of the area's transportation system. The plan must address at least a twenty year planning horizon.

The TIP is the region's spending plan for anticipated transportation improvement, containing a multi-year prioritized list of projects (3 years at a minimum) proposed for funding or approval by the Federal Highway Administration (FHWA) or the Federal Transit Administration (FTA). The TIP is the means for implementing the goals of the long-range transportation plans. Outside of metropolitan areas, planning is done by the state's department of transportation in concert with rural, non-metropolitan counties.

Before a TIP or a plan is adopted, the MPO and, subsequently, the U.S. Department of Transportation, are required to show that the area's planned transportation activities are consistent with (or "conform" to) the purpose of the SIP in nonattainment and maintenance areas; this process is known as the transportation conformity determination process.

The transportation conformity determination process links transportation and air quality planning. The transportation conformity process is designed to make sure that new investments in transportation infrastructure do not worsen air quality or interfere with the "purpose" of the SIP. The transportation plan and TIP must result in emissions consistent with those allowed in the SIP. To demonstrate that the transportation plan and TIP are in conformity, the MPO must show that the area's future emissions from highway and transit transportation activities meet the motor vehicle emissions budgets in the SIP (in cases where an area does not yet have a SIP in place, a different type of test is used). The U.S. Department of Transportation, including both the Federal Highway Administration (FHWA) and the Federal Transit Administration (FTA), must approve the MPO's conformity determination.

3.6 WHO IS INVOLVED IN THE TRANSPORTATION PLANNING PROCESS?

In developing transportation plans and TIPS, coordination and cooperation among multiple agencies–MPOs, state and local transportation and air quality agencies, offices of the U.S. Department of Transportation (FHWA and FTA), and EPA– must occur. This coordination and consultation process is called *interagency consultation*. State and local agencies establish formal interagency consultation procedures for all of the agencies involved, in accordance with the transportation conformity rules (40 CFR 93.105). In addition, pro-active efforts to encourage early and continuing public participation in decision making are required.

3.7 HOW ARE THE AIR QUALITY AND TRANSPORTATION PLANNING PROCESSES LINKED?

In the past, transportation plans were developed independently of the state's air quality planning process and were not required to consider the effects of transportation activities on air quality. Similarly, SIPs developed by air quality planners often failed to consider the feasibility of their plans with respect to the development of transportation infrastructure. Air quality specialists cannot guarantee a continuing reduction in motor vehicle-related pollutants if vehicle miles traveled continue to grow, and transportation specialists cannot improve mobility if the emissions budget for an area is created without regard to air quality impacts. The transportation conformity process ensures the integration of the air quality and transportation planning processes to ensure consistency between the emissions resulting from highway and transit transportation projects and the motor vehicle emissions budget in the SIP.

CHAPTER 4 LINKING LAND USE, AIR QUALITY, AND TRANSPORTATION PLANNING

4.1 WHAT IS THE RELATIONSHIP BETWEEN LAND USE, TRANSPORTATION, AND AIR QUALITY?

The physical characteristics and patterns of land development in a region, also known as the *urban form*, can affect air quality by influencing the travel mode choices citizens have available to them. Certain types of urban form necessitate the use of personal cars and trucks for travel. For example, when jobs and housing are far away from each other, and mass transit is not available, people are dependent on cars for daily travel. Urban forms that make automobile travel a necessity can contribute to air quality problems.

However, other options for urban form do exist. For example, development patterns that locate jobs, housing and recreation in closer proximity to each other, can mean shorter and fewer car and truck trips, thus reducing vehicle miles traveled (VMT) and likely reducing motor vehicle emissions. Other development patterns have the potential to improve or mitigate air quality problems by providing and promoting alternatives to vehicular travel, such as mass transit, walking, or biking.

4.2 IN WHAT WAYS DOES URBAN FORM IMPACT TRAVEL ACTIVITY?

Numerous studies have been conducted over the years to explore the connection between urban form and travel activity, and the weight of the empirical evidence suggests that characteristics of urban form can be important factors in reducing VMT and emissions.[13]

In a 1998 review of literature on the link between urban form and travel behavior, Apogee/ Hagler Bailly concluded that urban form can have a discernible effect on travel behavior.[14] The authors found that, based on both empirical studies and simulation (modeled) studies conducted primarily in the early to mid-1990's, evidence exists to suggest that changing the current patterns of development (generally low density, single use, and auto-oriented) can reduce vehicle travel and, hence, air pollutant emissions from vehicles. According to the authors, "even though changes in urban form may take years to occur, the best regional transportation models suggest that altering urban form can affect travel and emissions measurably within a time frame of 10 to 20 years."[15]

---

[13] Literature reviews performed by: Apogee/Hagler Bailly. 1998 *The Effects of Urban Form on Travel and Emissions: A Review and Synthesis of the Literature.* Prepared for U.S. Environmental Protection Agency, Urban and Economic Development Division, Washington, DC; and Johnston, R.A., Rodier, C. J., Choy, M., and Abraham, J.E. 2000. *Air Quality Impacts of Regional Land Use Policies.* Prepared for U.S. Environmental Protection Agency, Urban and Economic Development Division, Washington, DC.

[14] Apogee/Hagler Bailly. *The Effects of Urban Form on Travel and Emissions: A Review and Synthesis of the Literature.* 1998 Draft Report. Prepared for U.S. Environmental Protection Agency, Urban and Economic Development Division.

[15] Ib id. pg 56.

In the report, the authors note that there are a number of urban form features that can affect travel activity, the most significant of which are:

- Density,
- Land use mix,
- Transit accessibility,
- Pedestrian-environment/ urban design factors, and
- Regional patterns of development.

See Table 4 for a more complete description.

TABLE 4: FIVE CHARACTERISTICS OF URBAN FORM THAT INFLUENCE TRAVEL AND AIR QUALITY

DENSITY
- Refers to the compactness of a neighborhood, a development, or a region
- Can reduce vehicle travel by reducing the distances that people have to drive, reducing the necessity of owning a vehicle, and increasing the viability of using other modes of travel such as walking or biking
- Higher density development also makes mass transit more economically feasible for the public sector

LAND USE MIX
- Refers to incorporating different land uses (e.g., recreation, housing, employment, shopping) within a development, a neighborhood, or a region
- Can lead to shorter trip distances and greater use of walking, as well as a reduced need for vehicle ownership
- Can reduce required commute distances

TRANSIT ACCESSIBILITY
- Refers to locating high-density commercial and residential development around transit stations; also known as "transit oriented development," or TOD
- Can increase the market for such services, increase ridership, and decrease auto use
- Can lead to decreased auto ownership

PEDESTRIAN-ENVIRONMENT/URBAN DESIGN FACTORS
- Refers to features that improve the pedestrian environment such as sidewalks, clearly marked crosswalks, shade trees, benches, and landscaping; also refers to features that improve the bicycling environment such as bike paths and dedicated bike lanes, bike parking and clear signs
- Can reduce driving by increasing the desirability of walking and biking
- Can lead to decreased auto ownership

REGIONAL PATTERNS OF DEVELOPMENT
- Refers to patterns of dispersion, centralization or clustering of activities within a metropolitan area, as well as to the relationship of development to highway and transit systems; involves the interrelationships between employment and residential development and the transportation connection between sets of origin and destination points
- Can reduce driving by locating trip origins and destinations closer together
- Can encourage transit use and reduce vehicle trip making by concentrating development around transit networks and clustering development

Among other findings, Apogee/Hagler Bailly concluded that the weight of the evidence suggests that:

- Trip lengths can be reduced when compact development and integration of land uses occurs, even where automobiles are the dominant mode of transportation;

- Increased accessibility to multiple uses reduce average trip lengths;

- Urban form can have a measurable impact on the desirability of using different modes of transportation;

- Rates of vehicle ownership are lower in places where personal vehicles are not required for personal mobility, even when income/economic factors are considered;

- Accessibility to a variety of trip purposes, as in mixed use developments, may induce additional trips; however, these trips are shorter and are more likely to be made by walking than trips in areas where mixed land uses are not available; and

- Synergies between different land use factors can be important in influencing travel behavior, and changing one single factor may not be enough to change travel behavior.

All of these considerations suggest that decisions about urban form have the potential to influence such problems as traffic congestion, sprawl, air pollution, and other environmental and social conditions that are important to communities.

Numerous studies have attempted to quantify the impacts of urban form on VMT and emissions. Table 5 provides some examples of estimated impact; however, it is important to note that due to the many variations between communities, locations, and project characteristics, it is not possible to provide a simplified table of emissions reductions. Each land use activity should be examined individually to assess the impact it may have on VMT and emissions.

TABLE 5: ESTIMATING THE IMPACTS OF LAND USE ACTIVITIES ON VMT AND EMISSIONS— A SAMPLE OF STUDIES

The air quality impacts of land use activities on transportation depend on numerous factors, including density and location of development, amount of development, mix of uses, and access to transportation alternatives. The interaction of these factors is complex, and, due to the variations from one development project to another, each development should be analyzed individually. However, the results of the following analyses give some indication of the potential for VMT and emission reductions from land use activities.

- In **Portland, Oregon,** modified travel demand models, developed as part of the Making the Land Use, Transportation, Air Quality Connection (LUTRAQ) demonstration project, were used to compare a base case scenario of extending highway and road capacity to a scenario which emphasized transit-oriented developments as well as pedestrian/bicycle improvements and transportation policies (the LUTRAQ alternative). Daily VMT was shown to decrease by 8% for the LUTRAQ alternative, and, although congestion was expected to increase in this scenario, overall air quality was expected to improve, with NOx emissions expected to decrease by 6% and CO emissions decrease by 3%.[16]

- Researchers at the University of California and the University of Calgary performed simulation studies to estimate the regional impacts of land use activities.[17] The authors conclude, based on an study they conducted in **Sacramento, California**, that land use and transit policies may reduce VMT and vehicle emissions by approximately 4-7% over a 20 year time horizon.

- A 1994 study conducted by Cambridge Systematics which analyzed the mode of commute for workers in 330 companies in the **Los Angeles, California** region suggests that the presence of mixed land use can increase commuter trips using transit by 1.9%, while also encouraging bicycle trips.[18] The authors also found that the combination of mixed land uses and incentives for using transit resulted in a 3.5% increase in commute trips by transit.

- In **Baltimore, Maryland,** simulation studies suggest that a centralized development pattern would generate 0.9 % less daily VMT and 1.7% less severely congested VMT than the current development pattern over a period of 20 years. A decentralized development pattern was estimated to *increase* VMT by 1.8% daily and 1.6% in severely congested areas.[19]

- In **Washington, DC**, modeling of a jobs/housing balance scenario in the region predicted that transit use would increase and average trip length would decrease, due to greater proximity of housing to jobs, and vehicle trips per household would decrease by 5%. When combined with additional transit accessibility, a VMT reduction of 9.2% was predicted.[20]

---

[16] Cambridge Systematics, Inc. and Parsons, Brinckerhoff, Quade & Douglas. 1996a. *Making the Land Use Transportation Air Quality Connection: Analysis of Alternatives*. Vol. 5. Prepared for Thousand Friends of Oregon.

[17] Johnston, R.A., Rodier, C. J., Choy, M., and Abraham, J.E. 2000. *Air Quality Impacts of Regional Land Use Policies.* Prepared for U.S. Environmental Protection Agency, Urban and Economic Development Division, Washington, DC.

[18] Cambridge Systematics. 1994. *The Effects of Land Use and Travel Demand Strategies on Commuting Behavior.* Prepared for the U.S. Department of Transportation, Federal Highway Administration, Washington, DC.

[19] DeCorla-Souza, P. 1992. "The Impacts of Alternative Urban Development Patterns of Highway System Performance." Presented at ITE conference on Transportation Engineering in a New Era.

4.3 HOW CAN URBAN FORM BE CHANGED TO IMPROVE AIR QUALITY?

Communities make decisions about how best to direct future growth as part of their land use planning process. Across the country, an increasing number of communities are spreading further and further away from urban cores into low-density, car-dependent communities with housing, employment, shopping and recreation in separate locations. This suburbanization development pattern, sometimes referred to as sprawl, has been linked to a variety of community concerns, such as loss of open space, air and water pollution, fractured neighborhoods and traffic congestion.

Some communities are beginning to consider ways to accommodate future growth that take into consideration not just economic concerns, but environmental and quality of life concerns as well. As described in the previous section, strategies for land development that encourage higher densities, a mixture of land uses, greater accessibility to transit, development of pedestrian/bicycle infrastructure, and consideration of broader regional development patterns can slow or curb the potentially adverse effect that sprawl can have on air quality. Some examples of land use strategies include:

- **Transit-oriented development (TOD):** encouraging transit travel by developing moderate- to high-density housing, shopping, and employment centers along a regional transit system, with pedestrian access.

- **Infill development:** encouraging pedestrian and transit travel by locating new development in already developed areas, so that activities are closer together.

- **Brownfield redevelopment:** remediation and redevelopment of under-utilized or abandoned lands, usually in already developed areas, that have been contaminated during previous use.

- **Mixed-use development:** development that locates complementary land uses such as housing, retail, office, services, and public facilities within walking distance of each other.

- **Neotraditional design/pedestrian-oriented development:** creating a combination of land development and urban design elements with the purpose of creating pedestrian oriented neighborhoods.

- **Developing concentrated activity centers:** encouraging pedestrian and transit travel by creating "nodes" of high density mixed development, that can be more easily linked by a transit network.

- **Strengthening downtowns:** encouraging pedestrian and transit travel by making central business districts concentrated activity centers that can be the focal point for a regional transit system.

- **Jobs/housing balance:** reducing the disparity between the number of residences and the number of employment opportunities available within a sub-region by directing employment developments to areas with housing, and vice versa.

Note that this is not an exhaustive list; other possible strategies exist and are being considered in various communities around the country.[21]

---

[20] DeCorla-Souza, P. 1992."The Impacts of Alternative Urban Development Patterns of Highway System Performance." Presented at ITE conference on Transportation Engineering in a New Era.

[21] See Appendix A for additional examples of land use strategies and policies.

LAND USE STRATEGIES IN ACTION: TWO CASE STUDIES

Portland, Oregon

The Portland area is well known for its commitment to "smart growth" strategies which are designed to curb sprawl and encourage use of walking, biking and transit. Pearl Court Apartments and Orenco Station offer two examples.

The Pearl Court Apartment Complex is a full-block housing development in Portland's growing River District. The development, covering 0.74 acres, was built on a remediated brownfield site which had been contaminated with petroleum, lead, and other toxic residues from years of use as a railroad yard. Due to the high level of contamination, the developer was able to buy the 40-acre parcel in 1991 for well below market value. The sale of this land was made feasible under a policy known a Prospective Purchaser Agreement, which protects the buyer and subsequent owners from liability for contamination from past activities.

The redevelopment efforts were completed in 1998, providing 199 high-density, energy efficient and affordable urban housing units with pedestrian access to bus and rail lines. In addition, the development has a bicycle storage room that can hold 144 bikes. Future plans include development of a trolley. Due to all of these features, the city reduced the required number of parking spaces the developer had to provide.

Orenco Station is a 190-acre master-planned community located in Hillsboro, Oregon, a suburb of Portland. The community includes 1,834 residential units, including apartments, single family houses, live/work townhouses, and condominiums in a wide range of sizes and costs. In addition, the plan includes a mixed use town center with ground level retail shopping and housing units above, and many pedestrian and transit-oriented features. The development is located within walking distance of the Orenco Station stop, which is on a light rail line that connects Portland to its suburbs. The project was specifically designed to encourage walking and community interaction through the town center.

Chattanooga, Tennessee

In 1969, Chattanooga, Tennessee was ranked as having the worst air quality in the United States. By 1988, the city was meeting federal air quality standards. This achievement came about in part due to planning and policy decisions that began with vision planning and ended with specific, achievable goals.

In the 1970's, business leaders, local government officials, and civic groups collaborated to create a vision plan to revitalize the community of Chattanooga, Tennessee and create a sustainable future. From this process, they identified 40 goals and developed hundreds of projects to achieve their goals. To date, more than 85% of the goals have been reached. Among the goals was the development of a sustainable transportation system, which included the creation of a mixed-use, pedestrian oriented downtown; renovation of an old automobile bridge into a pedestrian walkway over a river; re-direction of federal transportation funding from highway building to maintenance of existing roads and transit systems, and implementation of a system of hybrid electric- and gasoline-powered buses in the downtown area to reduce short automobile trips.

The city of Chattanooga, like many U.S. cities, was faced with the increasingly prevalent trend of retail development locating in outer-ring suburbs and in rural areas, resulting in the decline of their downtown area. Chattanooga provides an example of the kinds of development projects that can revitalize economic activity in downtown areas and encourage pedestrian activity that may curb automobile trips for shopping and recreation.

One example of Chattanooga's innovative revitalization projects is Warehouse Row, a 322,965 square foot project where eight historically significant but vacant old warehouses were converted into a complex that mixes retail shopping space with prime office space. Funding for the project came from a combination of historic preservation tax credits, urban development action grants, and industrial revenue bonds. In addition, city and county funding resulted in the creation of a one-acre public park. These funding incentives, along with appropriate zoning, have made the creation of a viable, mixed use downtown area possible.

These kinds of land development strategies are often characterized as "smart growth" strategies. The term "smart growth" has been used by a variety of groups, and many people interpret its meaning in different ways. However, in publications by groups as diverse as the EPA, the Urban Land Institute, and the National Association of Home Builders, there has been agreement that certain land use strategies, such as infill development, mixed use development, transit-oriented development, and higher density development are aspects of smart growth.[22]

In some states and municipalities, existing policies and regulations make smart growth strategies impossible to achieve, or discourage their implementation. Changes to existing policies or creation of new policies may be necessary to make smart growth possible.

4.4 WHAT KINDS OF ACTIONS ARE NEEDED TO MAKE "SMART GROWTH" STRATEGIES ACHIEVABLE?

There are many actions that all levels of government can take to encourage land use strategies that promote alternatives to single occupancy vehicle travel. Some of these are discussed below.

LOCAL GOVERNMENT ACTIONS

At the local level, actions generally fall into three categories: regulations such as zoning and subdivision regulations, monetary incentives, and non-monetary incentives.

Zoning
Zoning can be used to allow or require mixed use development. Special districts can be designated where development must meet specific requirements for mixing of housing, employment, shopping and public services. Local governments can also use zoning to increase density levels in downtown areas or in areas surrounding transit stations. Other zoning policies such as fine grain zoning (replacing large single use areas with smaller zones that can accommodate a mix of uses) and overlay zoning (adding a second use to an already-zoned area), as well as standards for street design (requiring narrower, better connected streets with sidewalks, bicycle lanes, and bus stops) provide local governments with the regulatory means to redirect urban form in their communities.

---

[22]Ruma, Charles. 1999. "Smart Growth: Building Better Places to Live, Work and Play." National Home Builders Association (NAHB), Washington, DC. Visit NAHB on the web at www.nahb.com .

David O'Neill. 1999. "Smart Growth: Myth and Fact." ISBN 0-87420-835-1. ULI-the Urban Land Institute, Washington, DC. Visit ULI on the web at www.uli.org.

EPA's Smart Growth Network document on Smart Growth Principles: http://www.smartgrowth.org/information/principles.html.

Monetary incentives
Another way that local governments can implement smart growth strategies is to give tax breaks to developers who build in desired locations. For example, local governments can encourage employers to locate near existing housing areas and near transit routes by offering tax incentives. In cities and counties where developers are required to pay impact fees (fees to pay for additional infrastructure needs that the new development generates), local governments can set those fees higher in outlying areas than in existing urban cores, thus potentially making urban development more economically feasible for the developer. Local and state governments can also partner with financial institutions to provide financial incentives to home buyers (such as reduced rate mortgages or financial credits toward home purchases), to encourage them to live closer to their employers or closer to transit.

Non-monetary incentives
Local governments can also provide non-monetary incentives, such as accelerated permit processing or reduced parking requirements to encourage developers to use smart growth principles.

The combination of several local regulatory- and incentive-based programs may increase the likelihood that development will occur in desired areas.

EXAMPLES OF LOCAL ACTIONS TO PROMOTE SMART GROWTH

Austin, Texas

The city of Austin, Texas has developed a tool called "Smart Growth Criteria Matrix," a scoring system to help city planners determine if a proposed development deserves financial support from the city. The matrix is used to evaluate criteria such as 1) the location of development; 2) proximity to mass transit; 3) pedestrian-friendly urban design characteristics; 4) compliance with nearby neighborhood plans; 5) increases in tax base; and other policy priorities. The tool allows a development to accrue as many as 635 points in 14 categories, with higher scores being allocated for projects that incorporate smart growth principles such as mixed residential, retail and office uses, human scale detailing, and higher density development. Development projects that score within certain ranges, thus demonstrating that they support the goals of the city, may receive economic incentives such as waiver of developments fees, coverage of utility charges, and investments in infrastructure construction.

The city also offers Smart Growth Zone Specific Incentives, which refer to reductions in fees the City charges for zoning, subdivision, and site plan applications, and for water and wastewater capital recovery fees for development that occurs in specific areas designated by the city as priority growth areas.

Wilmington, Delaware

During the development of their 1996 long-range transportation plan, the Wilmington Area Planning Council (WILMAPCO), the area's metropolitan planning organization, recognized the need to address the relationship between land use and transportation. As a pilot study, WILMAPCO worked with community leaders in Middletown, Delaware, and the Delaware Department of Transportation (DelDOT) to explore alternative strategies for growth.

While currently a small rural community, Middletown expects its population of 4,200 to triple in the next 20 years. Existing zoning codes in the town were destined to lead to a pattern of cul-de-sacs and street linkages that were unfriendly to pedestrians and bikes. As the community grows, such a development pattern would likely lead Middletown residents to drive more and further than they currently do.

In an effort to prevent this type of growth pattern, local government officials in Middletown worked with WILMAPCO to develop a new zoning code that allows mixing of neighborhood-scale commercial services with residential areas. The code, which employs many of the principles emphasized in neotraditional design, specifies smaller building setbacks to place buildings closer to the streets, thus encouraging pedestrian traffic in retail areas. Aspects of the old code such as restrictions on alleys were removed, and the code encourages developers to locate garages behind buildings.

WILMAPCO also worked with DelDOT and the Middletown Town Council to develop and adopt new street design standards that reduce speeds of cars, thus making pedestrian travel safer.

The coordination of the local government, the metropolitan planning organization, and the Department of Transportation has resulted in a set of local policies that will help this town retain its small-town character.

STATE GOVERNMENT ACTIONS

A recent report published by the National Governors' Association identified three core areas where states can positively impact land use planning decisions[23]. These are through state activities such as:

- **Leadership and public education strategies**
  - State-wide vision planning
  - Production and dissemination of cost/benefit information to the public
  - Planning tool development
  - Fostering public/private partnerships

- **Economic investment and financial incentive strategies**
  - Targeting state funds to support statewide development goals
  - Supporting and promoting brownfields redevelopment efforts
  - Using tax policy to encourage redevelopment of older cities

- **Government collaboration and planning strategies**
  - Developing statute and laws that foster state and local collaboration on land use planning
  - Reducing barriers to development in targeted areas

While these state-level activities do not directly alter local planning and decision making processes, they can indirectly encourage urban form patterns that decrease automobile use.

In addition to these strategies, which encourage development in urban cores and discourage development at the urban fringe, many states and local communities are reexamining existing policies and funding practices to make sure that they aren't inadvertently encouraging development in untargeted areas. Examples include limiting access to new highways and bypasses, and limiting state funding for infrastructure improvements outside of designated growth areas. Some states have designated growth boundaries, outside of which the state will not fund infrastructure improvements needed to support development. Other states have implemented policies such as purchasing development rights on rural lands to ensure that they remain undeveloped.

---

[23] Hirschhorn, Joel S. 2000. *Growing Pains: Quality of Life in the New Economy*. National Governors' Association, Washington, DC. For more information, visit the National Governors' Association web site at http://www.nga.org/Center.

EXAMPLES OF STATE-LEVEL ACTIONS TO PROMOTE SMART GROWTH

Missouri

In urban areas across the country, old, contaminated plots of land lie underutilized or abandoned. Redevelopment of these "brownfields" into residential and/or commercial uses could, over time, revitalize old urban cores, and reduce transportation related emissions from commuting because people can live closer to where they work. However, the cost of cleaning these sites and liability concerns from buyers often prohibit efforts to reclaim them. In Missouri, as in other states, policies are now being adopted to make redevelopment feasible and desirable.

The state of Missouri has developed a brownfield redevelopment program through its Department of Natural Resources and its Department of Economic Development. The Department of Natural Resources oversees owner development and site remediation through its Voluntary Cleanup Program. The program's core goal is to reduce remediation costs. The Department of Economic Development, in concert with this program, has developed a Brownfield Redevelopment Financial Incentive Program, which provides tax credits, loan guarantees or grants to developers willing to reclaim, clean up, and redevelop contaminated, abandoned properties.

Another pair of Missouri policies designed to encourage redevelopment is the Historic Preservation Program and the Historic Tax Credit Program. These programs provide preservationists and developers with incentives to rehabilitate old structures and make them available for reuse as residential or commercial properties.

These state-level policies provide incentives for voluntary efforts to reclaim and rebuild older cities in the state.

Pennsylvania

Several recent initiatives are underway to influence the way land is developed in the State of Pennsylvania. In 1998, the Governor signed legislation that creates geographic areas known as Keystone Opportunity Zones. These zones are "no tax" zones, where state and local governments agree to eliminate all state and local taxes for employers and residents. In addition, designated zones receive a one-time funding grant for planning activities necessary to prepare for new development, and other aid in the form of reduced-rate loans and access to other grants. The policy is designed to inspire economic development and job creation in areas that had previously been neglected, primarily urban core areas. Such redevelopment of housing and jobs can reduce the need for vehicle travel to other locations to meet those needs.

In late 1999, the Governor signed into law the Growing Greener initiative, which provides nearly $650 million in funding for a variety of local land-use projects. In another legislative action, the Growing Smarter initiative, the State has committed to providing new land use planning tools for local governments.

FEDERAL GOVERNMENT ACTIONS

In early 2000, the White House issued a report defining the Federal government's role in promoting livable communities.[24] The report focuses on four ways that the Federal government can support community growth and development: 1) providing financial or regulatory incentives, 2) providing information, tools, and technical assistance to enhance local decision-making, 3) ensuring that Federal policies for land development and building use are consistent with community goals, and 4) building partnerships with communities, regions, the private sector, non-profits, and academic institutions in places across the country. Some examples of Federal actions include development of models and tools, such as the Smart Growth Index model, which aid local decision makers in weighing different development options; creation of grant programs to fund local planning efforts; and development of partnerships and pilot programs with state and local governments to explore the impacts of smart growth policies.

EXAMPLES OF FEDERAL GOVERNMENT PROGRAMS TO HELP STATES AND LOCAL GOVERNMENTS PROMOTE SMART GROWTH

Transportation funding

The FY2000 Federal spending budget included $9.1 billion dollars for transportation related measures. Through programs like the Congestion Mitigation and Air Quality Improvement (CMAQ) Program and the Transportation Enhancements Program, the Federal government can provide support for state and local efforts to ease congestion and reduce air pollution by funding programs such as incentive programs for ridesharing, improved transit facilities, and creation of bicycle and pedestrian paths. Through the Transportation and Community and Systems Preservation Pilot (TCSP) program, grants are provided for state and local planning agencies and governments to encourage coordination of transportation and land use planning while considering economic development and environmental concerns. Additional funding was allocated in the Transportation Equity Act for the 21st Century (TEA-21) for the New Starts program, which is designed to provide financial support to locally planned, implemented and operated rail transit systems.

Brownfield redevelopment

The Brownfields Economic Redevelopment Initiative is an umbrella program for a number of support and funding programs. The Brownfields National Partnership, one of these programs, is a collaborative effort by more than 20 Federal agencies to provide financial and technical support for brownfield cleanup. The Brownfields Cleanup Revolving Loan Funds is another program which provides grants for state, tribal and local cleanup efforts. The Clean Air Brownfields Pilot program represents a partnership among the U.S. EPA, the Economic Development Administration, and the U.S. Conference of Mayors to test methods of quantifying the air quality and economic benefits of redeveloping brownfields in Dallas, Chicago, and Baltimore.

For government actions such as those described here to be effective, cooperation and coordination at all levels of government is essential. Adoption of regulatory policies can be a lengthy process, and for non-regulatory policies such as regional or state-level planning guidelines, much effort is often necessary to encourage local governments, developers, and citizens to implement them. Therefore, these actions should be part of long-term strategies, and may take some time (as much as 10 to 20 years) before they produce broad-scale changes in development patterns and urban form.

---

[24] A Report From the Clinton-Gore Administration: *Building Livable Communities: Sustaining Prosperity, Improving Quality of Life, Building a Sense of Community.* June, 2000.

SECTION 2: POLICY AND TECHNICAL CONSIDERATIONS- ACCOUNTING FOR THE AIR QUALITY BENEFITS OF LAND USE ACTIVITIES

Section 2 provides general policy and technical guidance to air quality and transportation planners on how emissions reductions from land use activities can be incorporated into the air quality and transportation planning processes.

Chapter 5   **Accounting for emission reductions from land use activities** .... -27-
     5.1   What kinds of land use activities can be accounted for in SIPs and conformity determinations? .... -27-
     5.2   How are land use activities incorporated into the air quality and transportation processes? .... -27-
         5.2.1   What are land use planning assumptions? .... -28-
         5.2.2   What is the travel demand forecasting process? .... -29-
         5.2.3   What is the emissions modeling process? .... -30-
     5.3   How are land use activities incorporated into the SIP and the conformity determination? .... -31-
     5.4   What are the ways that I can account for land use activities? .... -31-

Chapter 6   **Including land use activities in the initial forecast of future emissions in the SIP** .... -33-
     6.1   What is the initial forecast of future emissions? .... -33-
     6.2   When is an initial forecast of future emissions made? .... -33-
     6.3   How can I account for "smart growth" activities in the land use assumptions that are made for the SIP? .... -33-
     6.5   What is "double counting?" .... -37-
     6.6   What else should I consider when including land use activities in my initial forecast of future emissions? .... -38-

Chapter 7   **Including a land use activity as a control strategy in the SIP** .... -39-
     7.1   What is a control strategy? .... -39-
     7.2   When would I include land use activities as control strategies in the SIP? .... -39-
     7.3   How can I account for land use activities as control strategies in my SIP? .... -39-
     7.4   Land use activities as Traditional Control Strategies .... -39-
         7.4.1   What are Traditional Control Strategies? .... -40-
         7.4.2   What are the existing statutory requirements for including a land use activity as a traditional control strategy in a SIP? . -40-
         7.4.3   What happens if I have included a land use activity as a traditional control strategy in a SIP, and now I have information that the land use activity is not occurring? ..... -43-
         7.4.4   What else should I consider when including a land use activity as a traditional control strategy in a SIP? .... -43-
     7.5   Land use activities and the Voluntary Mobile Source Emission Reduction Program policy (VMEP policy) .... -44-

		7.5.1	What is the Voluntary Mobile Source Emission Reduction Program (VMEP) policy? . . . . -44-

- 7.5.1 What is the Voluntary Mobile Source Emission Reduction Program (VMEP) policy? . . . . -44-
- 7.5.2 What are the existing statutory requirements for including a land use activity as a VMEP control strategy in a SIP? . . . . -44-
- 7.5.3 When should I use the VMEP policy to include a land use activity in a SIP? . . . . -45-
- 7.5.4 What happens if I have included a land use activity as a VMEP control strategy in the SIP, and I now have information that it is not occurring? . . . . -46-
- 7.5.5 What else should I consider when including a land use activity as a VMEP control strategy in a SIP? . . . . -46-
- 7.6 Land Use Activities and the Economic Incentive Programs policy (EIP policy) -47-
  - 7.6.1 What is the Economic Incentive Programs (EIP) policy? . . . -47-
  - 7.6.2 How is the EIP policy related to the VMEP policy? . . . . -48-
  - 7.6.3 When should I use the EIP Policy to include a land use activity in a SIP? . . . . -49-
- 7.7 What steps are necessary for quantifying land use activities as traditional, VMEP, or EIP control strategies? . . . . -49-
- 7.8 Transportation Control Measures as control strategies . . . . -52-
  - 7.8.1 What are Transportation Control Measures? . . . . -52-
  - 7.8.2 How are Transportation Control Measures and land use activities related? . . . . -52-

Chapter 8 Including land use policies or projects in the conformity determination without having them in a SIP . . . . **-55-**

- 8.1 What is a conformity determination? . . . . -55-
- 8.2 How is conformity demonstrated? . . . . -55-
- 8.3 Does this guidance impose new requirements for including land use activities in a conformity determination? . . . . -56-
- 8.4 If I have included a land use activity in a SIP, does it have to be included in the conformity determination? . . . . -56-
- 8.5 Can I account for the emissions benefits of land use activities in a conformity determination without having them in a SIP? . . . . -56-
- 8.6 How are land use activities included in the conformity determination? . . . . -56-
- 8.7 What are the transportation conformity rule's requirements for land use assumptions? . . . . -57-
- 8.8 How are the land use assumptions in a conformity determination reviewed? . . . . -59-
- 8.9 What are control strategies? . . . . -59-
- 8.10 What are the conformity rule's requirements for control strategies? . . . . . . -60-
- 8.11 How do I determine whether a land use activity is a land use assumption or a control strategy? . . . . -60-
- 8.12 What are some examples of land use activities that fit in each category? . . . . -61-
- 8.13 What is "double counting?" . . . . -62-
- 8.14 What if a land use activity is too small to have an impact on the outcome of travel demand modeling? . . . . -63-
- 8.15 What if our area doesn't use a travel demand model for transportation planning? . . . . -63-
- 8.16 What are the advantages of accounting for land use activities in the conformity determination without having them in the SIP? . . . . -63-

Chapter 9 Additional considerations when accounting for land use activities in the SIP or the conformity process . . . . -65-
    9.1   How can I determine whether or not my land use activities might have air quality benefits? . . . . -65-
    9.2   How will the time frame for implementing the land use activities affect which accounting option I choose? . . . . -66-
    9.3   What other important issues should I be aware of in quantifying air quality benefits? . . . . -67-
        9.3.1   Accounting for interactions between land use activities . -67-
        9.3.2   Quantifying land use activities individually or as a group . . . . -67-
        9.3.3   Using conservative estimates . . . . -68-
        9.3.4   Taking into account the scale of the land use activity . . . . -68-
    9.4   How will EPA assist me with quantification? . . . . -68-

CHAPTER 5 ACCOUNTING FOR EMISSION REDUCTIONS FROM LAND USE ACTIVITIES

5.1 WHAT KINDS OF LAND USE ACTIVITIES CAN BE ACCOUNTED FOR IN SIPs AND CONFORMITY DETERMINATIONS?

Land use activities, in the context of this guidance, are actions initiated by local, regional or state governments, individuals, organizations, and developers that change urban form in ways that may decrease te use of motor vehicles and encourage alternative modes of transportation. Where these land use activities can be shown to reduce emissions from motor vehicles, they may be accounted for in the air quality and transportation planning processes.

Land use activities may be policies or projects.

Specific policies that do have the force of law and therefore could be considered for air quality accounting include zoning ordinances, subdivision regulations, parking codes, and development standards. These policies are generally adopted by local governments. Other specific policies may be incentive programs, which require voluntary participation by developers or citizens.

Specific Projects are generally site-specific, and usually occur on a relatively small scale, although large-scale developments, such as master planned communities, do occur. Usually, land use projects are initiated by private sector actions; however, partnerships and initiatives between government agencies, public advocacy organizations, and developers are often the catalyst for such development projects.

Some projects can be accounted for in air quality and transportation planning processes if their emission benefits can be quantified. Note that supporting transportation elements of land use developments, such as the addition of transit lines and stops in the area, may be accounted for as transportation control measures.

5.2 HOW ARE LAND USE ACTIVITIES INCORPORATED INTO THE AIR QUALITY AND TRANSPORTATION PROCESSES?

Land use activities are incorporated into the air quality and transportation processes by modeling the emission reduction impacts of land use activities.

Air quality planners must estimate future pollution levels from motor vehicles in their air quality planning process. Transportation planners are also required to estimate future motor vehicle emissions of the highway and transit projects in nonattainment and maintenance areas to meet the requirements of the transportation conformity process. The way land is developed–how residences, jobs, shopping, recreation, and other destinations are situated within an area–can impact the length and number of vehicle trips, as well as the speed at which they occur. Therefore, land use patterns can have an effect on travel activity and emissions.

To calculate the amount of pollution from motor vehicles, planners consider the ways that land will be used in the future and how the future transportation network will support those uses. In general, there are three steps to this process:

- Establish land use planning assumptions,

♦ Conduct travel demand forecasting, and

♦ Perform emissions modeling.

These three steps are described in sections 5.2.1 through 5.2.3.

5.2.1 What are land use planning assumptions?

Before a transportation planner can forecasts future travel and emissions, the planner must make some assumptions about the number of people that will live in each part the region, and the number and location of employment and other activities.

To establish these land use assumptions, transportation planners make forecasts about population, the economy, and land use:[25]

Population forecasts: To forecasts future population growth, planners examine current information about birth and death rates, and the rates of migration to and from the region.

Economic forecasts: Forecasting economic activity requires consideration of the population trends, the region's ability to attract and retain employers, and expectations about how these trends will change in the future.

Land use forecasts: Transportation planners look at a local areas' land use plans as they predict where new population and employment will locate within the region. These plans typically reflect the impact of policies and specific projects that are designed to direct expected growth into desired locations. Land use plans can simply incorporate current trends, or they can include policies and programs designed to change those trends. Examples include policies and projects that encourage redevelopment of urban cores, increase density, or protect open spaces from development.

These population, economic, and land use forecasts may be made by the transportation planning agency, or by the local governments or other planning agencies.

The information on population and employment may be generated using land use models, and/or determined by expert judgement. Some land use models include factors such as location of industrial and service employment, location of employee households, and comparative costs for land.

Land use forecasts are usually performed at the district level, and are then sub-allocated to transportation analysis zones. The land use forecast are then used as inputs into the transportation modeling, or "travel demand forecasting," process.

---

[25] Beimborn, E. A. "A Transportation Modeling Primer." Chapter in *Inside the Black Box: Making Transportation Models Work for Livable Communities*. Environmental Defense and Citizens for a Better Environment. 1996.

5.2.2 What is the travel demand forecasting process?

Travel demand forecasting is a process where transportation planners predict expected travel activity throughout a region. Travel demand models are commonly used to make these predictions through the "4-step modeling" process. This process is described briefly below.[26]

First, an area is divided up into travel analysis zones, from which trips originate and to which trips are destined. The amount of population, households and employment forecasted for a zone will affect how much travel will occur in and between zones. Employment is used to represent not only work activities, but also shopping, lunch and other types of trips. Then, the model performs the following 4 steps:

1. **Trip Generation:** The trip generation step uses the land use assumptions to estimate the number of trip ends (productions and attractions) for each zone. The trips are generated by trip type, such as "home-based work," "home-based other" or "non-home based."

2. **Trip Distribution:** The trip distribution step links the productions with the attractions. Demand for travel between two zones is related to the number of trips in and out of the zone, and the amount of impedance (i.e., the effect of time, distance, and/or cost on travel activity).

3. **Modal Choice:** In some areas, the travel demand model also produces estimates of trips by mode (e.g., highway, transit, or other modes). Mode choice models may take into consideration factors such as demographic group, cost, trip purpose, and relative travel times.

4. **Trip Assignment:** Trip assignment involves assigning vehicle trips to specific links of the travel network. Travel demand models also estimate the speeds that vehicles travel, based on how congested the road network is.

This process generates VMT and speed data that is directly used to estimate motor vehicle emissions.

---

[26] U.S. EPA. 1997. *Evaluation of Modeling Tools for Assessing Land Use Policies and Strategies.* EPA Report EPA420-R-97-007. USEPA Office of Air and Radiation, Office of Mobile Sources, Ann Arbor, MI.

5.2.3 What is the emissions modeling process?

The last step in the process is to model the emissions produced. Emission rate models are used to estimate emissions for the area, taking into consideration factors such as the mix of vehicles types, temperature, etc. These rates are then applied to the VMT and speed estimates from the travel demand model to calculate motor vehicle emissions rates.

5.3 HOW ARE LAND USE ACTIVITIES INCORPORATED INTO THE SIP AND THE CONFORMITY DETERMINATION?

Using the three steps described in sections 5.2.1 through 5.2.3, air quality and transportation planners can model the emissions reduction impacts of land use activities. The output for each step is the input for the next step.

Figure 3 shows the relationship between the land use inputs, travel demand forecasting, and the emissions modeling.

FIGURE 3. RELATIONSHIP BETWEEN LAND USE MODELING, TRANSPORTATION PLANNING, AND EMISSIONS MODELING.

Land Use Modeling (or expert judgement)	Travel demand forecasting	Emissions modeling
Output: Location of population and employment	Output: Vehicle miles traveled and speed of traffic	Output: Regional emissions rates for motor vehicles emissions

When air quality planners prepare SIPs, they first perform an initial analysis of future emissions to assess what emissions are expected to look like given what land use and transportation activities are planned at the time of the analysis. To create this "initial forecast of future emissions," *land use assumptions* are made, which are fed into transportation models, and the results of this modeling are input into an emissions factor model. An analysis of the impact of *control strategies* designed to reduce emissions over time is then performed. Control strategies may include land use activities that can reduce emissions from transportation.

When transportation planners perform a regional emissions analysis for a conformity determination, they must demonstrate that the impacts of new transportation activities do not create air quality problems for nonattainment and maintenance areas. In this analysis, the emissions from the future transportation activities are compared to either the SIP's emissions budgets, or when budgets aren't available, either emissions from the "no-build" scenario, and/or emissions from a prior year (the specific requirements depend on the pollutant and the area's classification). For the regional emissions analysis, transportation planners also use *land use assumptions* to create the forecasts, and, the conformity determination may include new *control strategies* not found in the SIP, which may be used to offset the emissions impacts of new transportation activities.

In summary, land use activities are incorporated into SIPs or conformity determinations as either land use assumptions or control strategies.

5.4 WHAT ARE THE WAYS THAT I CAN ACCOUNT FOR LAND USE ACTIVITIES?

There are three general ways that you can account for land use activities. These are:

- Including land use activities in the initial forecast of future emissions in the SIP.

- Including land use activities as control strategies in the SIP.

- Including land use activities in the conformity determination without including them in the SIP.

There are some similarities when accounting for land use activities in either the SIP process or in the conformity process. However, existing regulatory specifications for SIPs and conformity determinations will determine exactly how you must account for these activities in each process. These specifications, and their applicability to land use activities, are discussed in chapters 6, 7, and 8.

CHAPTER 6 INCLUDING LAND USE ACTIVITIES IN THE INITIAL FORECAST OF FUTURE EMISSIONS IN THE SIP

6.1 WHAT IS THE INITIAL FORECAST OF FUTURE EMISSIONS?

All control strategy SIPs and maintenance plan SIPs must have an inventory of current emissions, and a forecast of future emissions. The initial forecast of future emissions is the level of emissions in the future target year that will result if no additional control strategies are implemented. The initial forecast includes effects of existing Federal regulations or programs that will come into effect by the forecast year (for example, Federal regulations such as new emissions standards), but does not include effects of any additional explicit control strategies that are included in the SIP to improve air quality.

The motor vehicle portion of the initial forecast is based on modeling the transportation network that will exist by the forecast year. The first step in modeling the transportation network is to make *land use assumptions* for your area. When creating land use assumptions for your area, you should make sure that you take into account the effects that "smart growth" policies and projects will have on those assumptions.

6.2 WHEN IS AN INITIAL FORECAST OF FUTURE EMISSIONS MADE?

The initial forecast of future emissions is made when an area prepares a SIP for the first time, or performs a SIP revision. Therefore, if your area is not in the process of developing or revising a SIP, you would not have this option available to account for your area's land use activities. Instead, you may wish to consider accounting for your land use activities in your next conformity determination.

6.3 HOW CAN I ACCOUNT FOR "SMART GROWTH" ACTIVITIES IN THE LAND USE ASSUMPTIONS THAT ARE MADE FOR THE SIP?

Land use assumptions – the location of households and employment – are the beginning of the air quality modeling process. Some areas employ land use models to estimate what future land use will be, while other areas use the best judgment of planners. Although it is not possible to predict exactly what will happen in terms of future land use, the land use assumptions made in the SIP **must be based on the best available information** and **must be realistic** about what will happen in the future.

EPA examines the assumptions made for the initial forecast of future emissions to ensure that they are reasonable. In particular, EPA compares the SIP's forecasting assumptions to those made in the past. Typically, if a SIP is submitted with land use assumptions that are based on past trends, EPA is likely to believe these assumptions are reasonable. However, if EPA receives a SIP with land use assumptions that are radically different from previous assumptions, EPA will closely scrutinize these assumptions and look for a justification of why the assumptions are the best available and reasonable. Therefore, when submitting a SIP which includes land use assumptions based on **general land use trends**, it is important for you to carefully consider the basis for your land use assumptions and ensure that they are reasonable. Additional documentation from state and local agencies may be necessary in some cases. Initial forecasts based on inappropriate assumptions may not ultimately be approved.

To determine whether or not the land use assumptions are reasonable, EPA considers the following questions:

- Are the future land use trends plausible?

- Are the land use assumptions made very different from the land use assumptions used in previous SIPs or the last conformity determination?

- If so, are there reasons for the change?

- Is the change of a reasonable magnitude?

- How realistic are the future assumptions, given what kinds of development are currently happening?

- If dramatic changes are predicted, are there legal mechanisms in place to ensure the projected assumptions will in fact occur?

> **ILLUSTRATION: THE CHICAGO AREA TRANSPORTATION PLAN**
>
> In Chicago, land use forecasting is done by the Northeastern Illinois Planning Commission (NIPC), who give forecasts to the Chicago MPO and Illinois air quality planning agency. Chicago's most recent SIP and transportation plan conformity determination included assumptions that "past trends of decentralized land use would be moderated" – that is, there would be increased infill in the central part of Chicago. NIPC made these assumptions based on their judgement that the actions already underway or likely to be implemented will contribute to substantial reinvestment in existing communities and increased redevelopment will continue to occur.
>
> NIPC documented the kinds of policy tools that they expected would become widespread during the forecast period, which include policies to provide funding for infrastructure that would make infill and brownfield redevelopment more feasible; increased focus at the state and federal levels on funding efforts to promote economic development in older communities; tax credits for rehabilitation of older and historic buildings; and priority funding to maintain the existing transportation system. NIPC also gathered information about local government policies, and included the impacts of these policies in the forecasts.
>
> Using expert judgement, NIPC concluded that
>
> > "actions already underway or likely to be implemented will contribute to (1) substantial investment within existing communities, (2) increased redevelopment in communities which have experienced disinvestment, and (3) high standards of new development in areas where it can be accomplished in a cost-effective manner."
>
> These land use assumptions were used to prepare the region's SIP. When the documentation on the planning assumptions was submitted to EPA Region 5, the region evaluated the assumptions, and determined the assumptions to be reasonable. The basis of this finding was that, although the assumptions were different from past trends, sufficient supporting evidence, including the current implementation of policies and the level of infill development already underway, indicated that a new trend was beginning and state and local policy and planning goals could realistically lead to the population, housing, and employment assumptions made in the plan.

In contrast to the discussion above, which is relevant to **general land use trends**, EPA believes that **specific policies and projects** should be included in an initial forecast of future emissions of a SIP under certain conditions, described below.

EPA believes that it would be appropriate to include a *specific land use policy* in the land use assumptions made for the initial forecast only if:

A. The policy meets one of the following conditions:
    - it has already been adopted by an appropriate jurisdiction, **or**
    - the policy is planned and there is an enforcing mechanism to ensure it will happen;

-and-

B. The effects of the policy haven't already been accounted for in the land use assumptions – that is, you are not double counting (this point is discussed further in section 6.5).

For example, suppose an area has passed a planning statute that requires local governments to establish an urban growth boundary. Because this is an adopted law, the effects of the urban growth boundary could be included in the initial forecast of future emissions. However, if an area is currently discussing whether to adopt an urban growth boundary, or one has been proposed but it is not yet adopted by an enforcing agency, it would not be appropriate to include its effects in the initial forecast. The urban growth boundary should be adopted before it is included.

ILLUSTRATION: THE MARYLAND SMART GROWTH POLICIES

In 1998, the Governor of Maryland signed an executive order establishing the Smart Growth and Neighborhood Conservation Policy, which implements the 1997 Smart Growth Areas Act. The cornerstone of this Act is the designation of "priority funding areas," or PFAs. These PFAs are areas where state and local governments have agreed that future growth and development should occur. The Act prohibits state agencies from funding or supporting infrastructure, economic development, housing and other programmatic investments outside of these designated areas. Other components direct state agencies to a) give priority to central business districts, downtown cores, and empowerment zones when funding infrastructure projects or locating new facilities; b) locate workshops, conferences and other meetings in these zones; and c) work with rural local governments to retain the rural character of their communities.

Maryland has four other complementary policies and programs. The Voluntary Clean Up and Brownfields program limits liability for developers of brownfield sites; The Live Near Your Work program, which provides home buyers with a minimum of $3,000 towards the home buying cost; the Job Creation Tax credits, which provides income tax credits to businesses that provide a minimum of 25 jobs within PFAs; and the Rural Legacy Areas program, which aims to preserve agricultural, forest and natural resource lands and protect them from development. The purpose of these incentives and programs is to complement the regulatory PFA policy by encouraging developers, employers, and home buyers to locate within the PFAs.

These policies are adopted at the State level, and State and local governments have worked together to designate PFAs. Therefore, it would be reasonable for Maryland to estimate the impacts of the PFA policy and the complementary incentive programs on the location of future population and employment and fold them into the land use assumptions in their initial forecast of future emissions.

EPA believes that it would be appropriate to include a *specific land use project* in the initial forecast of future emissions over and above the general assumptions only if:

A. The project meets one of the following conditions:
- it is already built,
- it is currently under construction, **or**
- it is planned, local zoning necessary for the project is already in place, and there is an enforceable mechanism to ensure that it will actually occur;

-and-

B. The effects of the project haven't already been accounted for in the general land use assumptions – that is, you're not double counting.

For example, suppose a large brownfield site near a transit line is currently being redeveloped as a mixed use, transit-oriented development that is designed to house and employ thousands of people. If the new population and new employment haven't already been accounted for, then this project can be included in the initial forecast of future emissions.

ILLUSTRATION: WASHINGTON'S LANDING, PITTSBURGH, PENNSYLVANIA

Washington's Landing is a brownfield revitalization project in Pittsburgh, PA. The redevelopment project is located on an island in the Allegheny River on a site that was once a stockyard and slaughterhouse. A two year environmental clean-up effort was required. The developer, Montgomery and Rust, its builder/ partner, the Rubinoff Company, and Pittsburgh's Urban Redevelopment Authority worked together to turn this underutilized site into a thriving community with townhomes close to downtown, a walk/bike path and a public park.

The development is primarily built, with 65 townhomes already sold and plans to build 23 more. As long as the population and housing growth has not been assumed already in some other way in the initial forecast of future emissions, the State of Pennsylvania could account for the emissions reduction impact of locating new growth in this infill/ brownfield location in their land use assumptions.

6.4 WHAT IS "DOUBLE COUNTING?"

EPA wants to ensure that effects of land use activities are not counted twice. Areas must be sure that what they are including in the initial forecast has not already been included in some other way. An area should include either the effects of a land use policy, or the effects of individual projects that happen as a result of that policy, but shouldn't count the effects twice.

For example, suppose a metropolitan region adopts a policy to give incentives to developers for building infill development downtown. Forecasts could be made on the amount and location of population and employment in the zones that would be affected by the policy. The state could then account for the impact of this policy in the land use assumptions for the SIP. However, once that is done, it would not be appropriate to also assume that new population and employment would occur for the individual developments that occur as a result of that policy. That would be doubling counting, because the new population and employment that result from the individual projects would have already been accounted for when the policy was included in the initial forecast of future emissions.

Likewise, if you have already accounted for the impacts of a large-scale new infill development on population and employment in the land use assumptions, it would not be appropriate to also account for the impacts of the infill incentive policy that caused the specific development to occur. Either the effects of the development or the effects of the policy should be counted, but not both.

6.5 WHAT ELSE SHOULD I CONSIDER WHEN INCLUDING LAND USE ACTIVITIES IN MY INITIAL FORECAST OF FUTURE EMISSIONS?

This option allows you to account for all of the smart growth policies, programs and projects that you are already doing. The composite impact of these smart growth activities may reduce your forecasted emissions level in the future, thereby reducing the amount of additional emissions reductions needed from control strategies.

Also, by associating air quality benefits with your smart growth programs on air quality, this analysis may be useful in your efforts to promote these programs more broadly. However, since this analysis is designed to set a baseline level of emissions, specific impacts of individual activities are not reflected in this analysis. States may want to demonstrate specific reductions associated with certain activities, and may wish to compute these impacts separately.

Also, because of the nature of the travel demand forecasting process, the effects of microscale design features are not well represented in this modeling process. Adjustments to the regional scale travel demand forecasting process may be necessary to capture the effects of microscale activities. This topic is discussed in greater detail in chapter 10.

Finally, it is important to note that inclusion of land use policies, programs and projects that differ greatly from past trends will be scrutinized for reasonableness, and may not be accepted as land use assumptions without additional justification (e.g., adopted commitments by implementing parties in place). Therefore, it is to your benefit that your analysis include support for your assumptions about the effects your land use policies and programs will have on future development patterns.

CHAPTER 7 INCLUDING A LAND USE ACTIVITY AS A CONTROL STRATEGY IN THE SIP

7.1 WHAT IS A CONTROL STRATEGY?

A control strategy is a policy, program, or project used by a nonattainment or maintenance area to reduce ambient air pollution levels in order to satisfy Clean Air Act requirements, such as attaining the standards, demonstrating reasonable progress towards attainment, or maintaining the standards. Control strategies are listed and described in the State Implementation Plan (SIP). Collectively, all of the control strategies in the SIP must reduce emissions enough to show attainment, maintenance, or further progress, depending on the type of SIP.[27]

Some examples of control strategies include use of clean fuels or advanced control technology to reduce emissions.

7.2 WHEN WOULD I INCLUDE LAND USE ACTIVITIES AS CONTROL STRATEGIES IN THE SIP?

There are two situations where you can include land use activities as control strategies in your SIP.

- If you are in the initial stages of preparing your SIP, you can include specific land use activities with all of your other control strategies which reduce emissions, or

- If you have already submitted a SIP, but discover that you need additional reductions, you can do a SIP revision.

Since it takes some time to develop a SIP, and there are very specific requirements that must be met, you should coordinate with your local metropolitan planning organization, state or local air quality agency, and EPA regional office as soon as you determine that you want to consider a land use activity as a control strategy in the SIP. Early coordination will help to make your land use control strategy successful.

7.3 HOW CAN I ACCOUNT FOR LAND USE ACTIVITIES AS CONTROL STRATEGIES IN MY SIP?

There are three ways that land use activities can be accounted for as control strategies in the SIP.

A specific land use activity that can be shown to have air quality benefits can be considered a control strategy in a SIP if certain requirements, outlined in the Clean Air Act, are met. For the purposes of this guidance, we will call these control strategies "**traditional control strategies.**"

---

[27] See 40 CFR Part 52.

In addition, EPA has developed two specific SIP policies that address innovative, voluntary control measures, which may also be applicable when accounting for land use activities as control strategies in a SIP. They are:

- **Voluntary Mobile Source Emission Reduction Program policy (VMEP policy),**

-and-

- **Economic Incentive Programs policy (EIP policy).**

Sections 7.4, 7.5, and 7.6 provide more details on these policies and will help you determine which control strategy option is best for you. Section 7.7 discusses quantification considerations. Lastly, transportation controls measures (TCMs) and their relationship to land use activities are discussed in section 7.8.

7.4 LAND USE ACTIVITIES AS TRADITIONAL CONTROL STRATEGIES

7.4.1 WHAT ARE TRADITIONAL CONTROL STRATEGIES?

Traditional control strategies, in the context of this guidance, are control strategies that can be shown to have air quality benefits and meet certain requirements outlined in the Clean Air Act. Land use activities that meet the existing statutory requirements of the Clean Air Act can be accounted for as traditional control strategies.

7.4.2 What are the existing statutory requirements for including a land use activity as a traditional control strategy in a SIP?

Under the Clean Air Act, traditional control strategies must a) be consistent with the purpose of the SIP and b) must not interfere with other requirements of the Clean Air Act. In addition, in order to be approved by EPA in a SIP, the emission reduction must be:

- **Quantifiable**
2. **Surplus**
3. **Enforceable**
4. **Permanent**
5. **Adequately Supported**

To include a land use activity as a traditional control strategy in a SIP, these five requirements must be met. These five requirements are described in further detail below.

Quantifiable

The emission reductions from the control strategy must be reliably calculated. Your emission reduction calculation methodology should be logical, supportable, and as thorough as reasonably possible. To demonstrate that this requirement has been met, you should provide:

- A complete narrative description of the land use activity, including the implementation schedule;

- A description of the applied emission calculation methodology and reasons for selection of that methodology;

- An estimate of anticipated emission reduction benefits, with documentation of assumptions and calculations;

- A discussion of quantification uncertainties;

- The data set used, and a description of the data collection methodology; and

- Other information, as applicable, that supports the reasonableness of the estimate.

Surplus

Emission reductions associated with the land use activity must not be relied upon in any other air quality program included in your SIP. In other words, you can not "double-count" your emissions. To demonstrate that this requirement has been met, you should provide:

- A statement that the appropriate agency has reviewed the control strategy and it is not accounted for in other parts of the SIP; and

- A statement describing the potential areas of overlap, if any, and steps to ensure that emission reductions are surplus and that there is no double-counting

Enforceable

Under the Clean Air Act, for any program to be considered a control strategy in a SIP, the actions required to achieve emissions reductions must be independently verifiable. Further, in the event that program violations occur (i.e., the implementation of the control strategy does not occur in the manner stated in the SIP submission), the state must identify those violations and the parties liable for the violations, and enforce the action and apply penalties where applicable.

To meet this requirement, the state must have the ability to enforce the control strategy. To demonstrate that this requirement has been met, you should provide:

- Evidence that a complete schedule to implement, and enforce the land use activity has been adopted by the implementing agency or agencies;

- A description of the monitoring program to assess the land use control strategy's effectiveness;

- Evidence that the land use activity was properly adopted by a jurisdiction with legal authority to commit to and execute the activity; and

- Evidence that the state has the ability to enforce the control strategy if violations occur.

This last item presents a special concern when considering land use activities as control strategies in SIPs. According to Section 110 (a)(2)(E) of the Clean Air Act, the state must provide "necessary assurances that the State (or, except where the Administrator deems inappropriate, the general purpose local government or governments, or a regional agency designated by the State or general purpose local governments for such purpose) will have adequate personnel, funding, and authority under State (and, as appropriate, local) law" to carry out the implementation of control strategies included in the SIP.

In the case of land use, states usually do not have enforcement authority over local land use decisions. To include a land use activity in a SIP as a traditional control strategy, the state must either have direct authority to enforce the activity, or the local government must have authority to enforce the activity.

Therefore, if a state wishes to account for a particular land use activity as a control strategy, and the control strategy involves an action being taken by the state itself, then the state can, of course, enforce the action. If, however, the control strategy involves an action which the state does not have direct enforcement authority over, such as a land use action taken by a local government or developer, **the state can only account for this land use action as a traditional control strategy IF there is an enforceable agreement in place**. This agreement may be in the form of:

1. An enforceable policy by a local government agency;

2. A commitment by a legal body (such as a regional agency) which can influence local decision making to ensure implementation; or

3. A memorandum of understanding among the appropriate parties (e.g., the state and the developer or the local government and the developer).

Without such a mechanism, the state cannot, under CAA rules, include a land use activity in a SIP as a traditional control strategy. However, there are two other SIP policies under which the land use activity may be considered a control strategy; these are discussed in section 7.5 and 7.6.

Permanent

To include control strategies in the SIP, the emission reductions associated with them must occur throughout the life of the control strategy, and for as long as it is relied upon in the SIP. To demonstrate that this requirement has been met, you should provide:

- Documentation showing that the control strategy will be implemented in a manner that ensures that the emission reductions will occur throughout the life of the SIP.

Adequately Supported

The state or responsible party must demonstrate that adequate personnel and program resources are committed to implement and enforce the program. To demonstrate that this requirement has been met, you should provide:

- Evidence that funding has been (or will be) obligated to implement the land use activity;

- Evidence that all necessary approvals have been obtained from all appropriate government entities (including state highway departments if applicable);

- Evidence of inclusion of the land use activity in a city/township/county development plan; and

♦ Other information, as applicable, that demonstrates adequate support.

7.4.3 What happens if I have included a land use activity as a traditional control strategy in a SIP, and now I have information that the land use activity is not occurring?

If you have already included a land use activity in a SIP, but now know that it is not occurring, or is not occurring in the way it was expected to when it was put in the SIP, you may not meet the goal of the SIP (further progress, attainment, or maintenance of the standard). Due to the enforceability requirements, the state may have to initiate enforcement procedures against the party responsible for implementing the control strategy. Also, in the next conformity determination you must be sure that you account for the land use activity as it is actually being implemented, rather than how it is described in the SIP.

7.4.4 What else should I consider when including a land use activity as a traditional control strategy in a SIP?

Including a land use activity as a control strategy in a SIP can help you meet your air quality goals by allowing you to account for emission reductions that you need to show attainment, progress, or maintenance. This may be an especially appealing option to areas that are having difficulty attaining, and are seeking all viable options for emissions reductions.

As a traditional control strategy, the activity will need to meet the criteria outlined in section 7.4.2 for EPA to approve it in the SIP. Having to meet these criteria can be an advantage or disadvantage depending on one's point of view. For example, the requirement that control strategies must be enforceable provides a guarantee that they will actually happen. However, a land use activity that is included as a traditional control strategy must be enforceable against the implementing party. For example, a local government may adopt a policy of high-density zoning in various areas in its jurisdiction. In many cities, while such a zoning policy is enforceable, the local government reserves the right to waive zoning requirements, in response to citizen complaints or developer requests. The need for enforceability, and/or funding at the local level to include such a zoning policy in a SIP could be a disincentive for some local governments to include their land use activities in the state's SIP, as it would, in effect, bind the local government to enforcing the action. The answer to this balance of incentive and disincentive must be addressed at the state and local level.

In cases where the state does not have regulatory authority to implement and enforce a land use activity against the source of the emissions, or where such an enforceable commitment from the implementing party cannot be obtained, the state may wish to account for the impacts of the land use activity under one of the two special EPA policies discussed in sections 7.5 and 7.6.

7.5 LAND USE ACTIVITIES AND THE VOLUNTARY MOBILE SOURCE EMISSION REDUCTION PROGRAM POLICY (VMEP POLICY)

7.5.1 What is the Voluntary Mobile Source Emission Reduction Program (VMEP) policy?

The Voluntary Mobile Source Emissions Reduction Programs policy, also known as the VMEP guidance or the "voluntary measures policy", was signed October 27, 1997[28]. This policy was developed to allow states to account for the benefits of voluntary mobile source measures in their SIPs. These are measures that rely on the voluntary actions of businesses or individuals to achieve emissions reductions. According to the policy,

> "Voluntary mobile source measures have the potential to contribute, in a cost-effective manner, emissions reductions needed for progress toward attainment and maintenance of the National Ambient Air Quality Standards (NAAQS). EPA believes that SIP credit is appropriate for voluntary mobile source measures when we have confidence that the measures can achieve emission reductions."

Some examples of voluntary measures, or VMEPs, include economic and market-based incentive programs, trip reduction programs, and growth management strategies.

7.5.2 What are the existing statutory requirements for including a land use activity as a VMEP control strategy in a SIP?

VMEP control strategies must meet the Clean Air Act's requirements that control strategies be quantifiable, surplus, permanent and adequately supported, as described in section 7.4.2. However, the VMEP policy defines the "enforceability" requirement differently than it is defined for traditional control strategies.

The common thread among all VMEPs is that, for these measures to be successful, voluntary participation of businesses and citizens is necessary. In other words, VMEPs are not required, and state and local governments do not have the authority to make participants take action.

Therefore, under the VMEP policy, the definition of enforceable for voluntary measures is different than for traditional control strategies. Voluntary measures that are included in SIPs are not enforceable against the source; rather, the state is responsible for ensuring that the emission reductions accounted for in the SIP do indeed occur. The state must make an enforceable commitment to monitor, assess and report on the emission reductions resulting from the voluntary measures and to remedy any shortfalls from forecasted emission reductions in a timely manner.

In general, for EPA to approve voluntary measures, a state must submit a SIP to EPA which:

- Identifies and describes the voluntary program;

- Contains projections of emission reductions attributable to the program, along with relevant technical support documentation;

---

[28] Memorandum to EPA Regional Administrators 1-10, from EPA Office of Mobile Sources: "Guidance on Incorporating Voluntary Mobile Source Emission Reduction Programs in State Implementation Plans (SIPs) October 24, 1998. Visit the Voluntary Measure web address at: http://www.epa.gov/oms/transp/traqvolm.htm.

- Commits to monitor, evaluate, and report the resulting emissions effect of the voluntary measure;

- Commits to remedy in a timely manner any SIP shortfall if the voluntary program does not achieve projected emission reductions; <u>and</u>

- Meets all other Clean Air Act requirements for SIP revisions.

Due to the innovative nature of voluntary measures, EPA's lack of experience in quantifying them, and the inability to enforce these measures against individual sources, EPA has set a limit on the amount of emission reductions allowed under the VMEP policy in a SIP. This limit is set at **three percent (3%)** of the total emission reductions needed to reach attainment.

For some land use activities, as you gain more knowledge about their impacts and are better able to quantify them, the three percent cap on VMEPs may eventually be limiting. EPA plans to review the three percent cap in 2002.

CREDITING INNOVATION AND EXPERIMENTATION:
Accounting for Land Use Activities using the Voluntary Mobile Sources Emission Reduction Programs Policy

The VMEP policy was designed to encourage innovation and creativity, and to give states a wider range of programmatic options to consider when developing strategies for reducing emissions from cars and other mobile sources. The types of programs being created under the VMEP policy attempt to gain additional emissions reductions beyond mandatory Clear Air Act programs by engaging the public to make changes in activities that will result in reducing mobile source emissions.

The programs rely on the *voluntary* actions of individuals or businesses to achieve emissions reductions; therefore, if these programs are included as VMEPs in a SIP, these actions are not directly enforceable against the party taking the action. In other words, the party being encouraged to take the action is not held responsible if they fail to take the action. The **state** must commit to remedy any emissions reduction shortfall in a timely manner if the VMEP policy does not achieve projected emission reductions.

7.5.3 When should I use the VMEP policy to include a land use activity in a SIP?

For land use activities where the state or local government does not have the authority to enforce the activity, a state could not include a land use activity as a *traditional control strategy* because they would not be able to meet the enforceability requirements stipulated in the Clean Air Act. In these cases, you may wish to account for the air quality benefits of the activity under the VMEP policy.

The VMEP policy allows for more innovation than most other EPA SIP programs. It is the appropriate policy to use for many newer programs that do not have well established quantification methodologies. In addition, by not requiring that the state directly enforce against a party implementing a measure, businesses and individuals may be more willing to try newer policies and programs.

> **ACCOUNTING FOR LAND USE ACTIVITIES AS CONTROL STRATEGIES UNDER THE VMEP**
>
> Civano, Arizona, is an 820-acre planned community located in Tucson, Arizona. The community evolved from a partnership between the private developers and community groups (Community of Civano, LLC, Trust for Sustainable Development and Case Enterprises) and state and local government agencies (the City of Tucson, the State of Arizona, the Arizona Energy Office, the Metropolitan Energy Commission, the State Land Trust). The goals of the project were to save energy, save water, reduce solid waste production and reduce air pollution. A core component of this community's development strategy was to design the community to minimize reliance on automobiles and encourage walking and biking. Design features include tree-lined biking and walking paths, narrow streets, lot layouts to encourage social interaction and to conserve open spaces, and a mix of housing, employment, and recreation within the development. The project estimates that, due to these design features, vehicle miles traveled within the development will be reduced 40% compared to traditional developments.
>
> The developers are *voluntarily* designing the site to decrease the need for driving, and the residents will *voluntarily* reduce their driving because of these features. Due to the fact that emission reductions are the result of ***voluntary actions*** of the developer and the residents, the State of Arizona could consider this development project a control strategy under the VMEP policy. Under this policy, the developer would be protected from any legal action in the event that the emissions reductions do not occur in the manner expected. In the event of a shortfall, it would be the State that is responsible for implementing other measures that make up the emissions shortfall.

7.5.4 What happens if I have included a land use activity as a VMEP control strategy in the SIP, and I now have information that it is not occurring?

As with traditional control strategies, if you have already include a land use activity in a SIP, but now know that it is not going to occur, or is not occurring in the way it was expected to when it was put in the SIP, you may not meet the goal of the SIP (further progress, attainment, or maintenance of the standard). Also, in the next conformity determination you must be sure that you account for the land use activity as it is actually being implemented, rather than how it is described in the SIP.

The biggest distinction between traditional control strategies and VMEPs is that, in the event a program does not achieve the emissions reductions predicted, the *state*, not the implementing party, is required to make up the shortfall.

7.5.5 What else should I consider when including a land use activity as a VMEP control strategy in a SIP?

Some actions taken by government agencies, such as regional policies to coordinate land use planning, and large-scale development projects, may require ten to twenty years before having any significant impact on emissions from motor vehicle sources. Since attainment SIPs are generally implemented over a shorter time frame than that, the air quality benefits of these land use activities may not be realized in the time frame covered by the applicable SIP. However, with careful planning, the benefits of land use activities adopted during the attainment planning period may be included in one (or both) of the two future maintenance planning periods, which last ten years each. For example, in a state where an ozone SIP must show attainment of the NAAQS in 2007, a local or state government incentive program to encourage development to locate in existing urban areas may not show emissions reductions in that time frame; however, the state could still benefit from the potential long-term benefits of this incentive program by accounting for impacts of the program in the subsequent maintenance periods.

In some cases, smaller scale development projects and policies, such as reduced fare transit passes or special mortgage programs for people purchasing homes in transit oriented communities, may have demonstrable benefits within the time frame of the attainment SIP and therefore may be suitable for inclusion in the SIP as VMEPs.

7.6 LAND USE ACTIVITIES AND THE ECONOMIC INCENTIVE PROGRAMS POLICY (EIP POLICY)

7.6.1 What is the Economic Incentive Programs (EIP) policy?

The Economic Incentive Programs (EIP) policy is a regulatory program designed to encourage the use of market-based incentives or information to reduce emissions. The purpose of the EIP policy is to provide flexibility in how sources meet their emission reductions targets, and empowers sources to find the most suitable and cost-effective means of meeting the goals of attainment or maintenance. The policy covers stationary, area, and mobile source emissions.

The EIP guidance document, "Guidance on Economic Incentive Programs for Air Quality," provides specific details on how to ensure that incentive programs meet the basic Clean Air Act criteria of quantifiable, enforceable, surplus, permanent, and adequately supported as listed in the traditional control strategy discussion[29]. In particular, the EIP guidance provides specific instruction on

- Creating and implementing an EIP;
- Including features to measure and track the results of the program;
- Evaluating results of the program;
- Including reconciliation procedures in the event that the EIP does not meet its predicted emission reduction goals;
- Ensuring the program meets SIP requirements for completeness and approvability; and
- Ensuring public participation.

The guidance describes four categories of EIPs: <u>financial mechanisms</u>, <u>public information programs</u>, <u>clean air investment funds</u>, and <u>trading programs</u>. At this time, financial mechanisms and public information programs are most applicable to including land use activities as EIPs in a SIP.

An example of a land use activity that could be an <u>financial mechanism</u> EIP might be a program that offers some type of monetary reward or incentive to developers. This incentive could be in many forms, including

---

[29] At press time, this guidance document was being redrafted and was not publicly available. The final version of the guidance is slated for release in January, 2001. The document will be posted on the web at http://www.epa.gov/ttn/oarpg/new.html and http://www.epa.gov/ttn/ecas/.

tax breaks, grants for brownfield redevelopment, or a flexible development fee structure. Alternatively, incentives could also be offered to home buyers or companies moving into areas where development is desired. These monetary incentives could be offered by local, state, regional, or federal government agencies, or perhaps even by non-governmental agencies. A public information program, such as a labeling program to designate environmentally sound or "smart growth" developments as being supported by the state could also be a land use EIP.

Generally, land use activities accounted for under the EIP policy would fall under the category of financial mechanisms. Due to the complexities involved in implementing both land use activities and emissions trading programs, EPA is not prepared to present specific guidance on how to incorporate land use policies into emissions trading programs at this time; however, we are continuing to evaluate the feasibility of incorporating land use activities into emissions trading programs, and seek comment on this issue.

EXAMPLES OF STATE FINANCIAL INCENTIVE PROGRAMS TO ENCOURAGE SMART GROWTH	
Kentucky:	Awarded $8 million in grant money to 21 Kentucky cities to revitalize downtown areas; requires 20% match by the cities
Michigan:	Created the Clean Michigan Initiative, which includes $335 million in funds to clean up of brownfields
Missouri:	Provides tax credits for home owners to encourage rehabilitation of older homes and construction of new homes in urban cores and established suburbs

7.6.2 How is the EIP policy related to the VMEP policy?

Economic incentive programs differ from voluntary measures in that under a state's EIP, emission reductions (or actions leading to emission reductions) **must either be identifiable and enforceable against a specific source or the state must use one of the following three methods to meet the enforceability requirement:**

- ♦ The EIP submittal includes fully adopted contingency measures and contains a state commitment to automatically implement contingency measures, if necessary;

- ♦ The state will only count emission reductions on a retrospective basis; or

- ♦ The state has used the control strategy in a similar situation, has achieved positive results, and gets preliminary approval from EPA to use the provision.

Some strategies might be originally approved under the voluntary measures policy and later, after program evaluations have been completed, could be approved as a regular EIP. If an emission reduction strategy can meet the EIP requirements, a state should strive for the strategy to be approved as an EIP rather than as a voluntary measure because EIP measures have a greater degree of certainty since they are more quantifiable than voluntary measures and are enforceable against the source. Also, EIP measures are not subject to a limitation, while voluntary measures are subject to a 3% cap.

The EIP guidance on developing and quantifying programs may also provide useful information for those wishing to account for land use activities with strong economic incentive components under the VMEP policy, due to the similarities of the policies.

7.6.3 When should I use the EIP Policy to include a land use activity in a SIP?

If your land use activity has a strong economic incentive component, such as including a fee structure for new development, and you want to include it in your area's SIP, you could include it as an EIP.

If you wish to use the EIP guidance to account for a land use activity, you should contact your EPA Regional Office. Since the EIP guidance is currently being redrafted, your regional EPA contact can assist you in determining whether your program will be consistent with the EIP policy.

7.7 WHAT STEPS ARE NECESSARY FOR QUANTIFYING LAND USE ACTIVITIES AS TRADITIONAL, VMEP, OR EIP CONTROL STRATEGIES?

There are five general steps you should follow to quantify land use activities in a SIP. They are:

1. Completely describe the land use activity.

2. Describe how employment and housing and other infrastructure are affected by the activity.

3. Describe how you determined the travel activity effects arising from the land use activity.

4. Determine the emissions reduction associated with the travel activity effects.

5. Demonstrate that the emissions reductions are surplus.

Step 1: Completely describe the land use activity

You should describe the land use activity as completely as possible. You should describe what the activity is, the actual physical changes that will result, population effects, zoning, density, the goals of the land use activity, travel activity effects and emission reduction potential. If you are including the land use activity as a traditional, VMEP or EIP control strategy, you will need to meet all of the specific documentation requirements specified for that policy.

Step 2: Describe how employment and housing and other infrastructure are affected by the activity

You should include in your analysis, an economic forecast, employment location, and household location, as well as any other pertinent infrastructure information. You should use accepted land use models. Check with your Regional EPA or DOT office for assistance in determining a model appropriate to your area. Many areas use expert judgement to estimate future land use and/or to determine the assumptions that underlay modeling. EPA will accept the use of expert judgement for land use assumptions, however you may want to adjust your estimates to reflect the confidence you have in the assumptions.

Step 3: Describe how you determined the travel activity effects arising from the land use activity

You should describe your methodology as completely as possible. Information normally used in transportation planning analysis or conformity analysis should be utilized where it exists so that all assumptions are consistent.

At this time, EPA generally recommends that you use existing travel demand models to analyze the impacts of your land use activities. However, it may be possible to modify your travel demand models to better capture the effects of land use activities. In the study, *Making the Land Use Transportation Air Quality Connection* (LUTRAQ), researchers modified existing land use and travel demand models to improve their ability to forecast the effects of development density and pedestrian-oriented design on travel behavior.[30] You should consult your local regional EPA or DOT office to determine which models are commonly used or applicable to your circumstances, and to obtain technical assistance when considering model modifications for your area..

Step 4: Determine the emissions reductions associated with the travel activity effects

You should always use the latest emissions model approved by EPA (the latest EMFAC modeling in California, and the latest MOBILE model for all other states).[31] You should always use the best data available when running the MOBILE or EMFAC model. Local data should be used when available. For vehicles and/or equipment (e.g. alternative fueled vehicles) not represented in these emission models you should use reputable data from actual testing programs.

Step 5: Demonstrate that the emission reductions are surplus

You must make sure that your quantification methodology only counts reductions that are surplus. You <u>cannot</u> account for actions or plans that already are included in your forecast of future emissions for your SIP or conformity analysis. To do so would be double-counting.

Emission reductions used to meet air quality requirements are surplus as long as they are not otherwise relied on in your SIP or other actions already required and accounted for under Federal rules that focus on reducing criteria pollutants or their precursors.

You will need to coordinate closely with other program offices in your air quality planning section to ensure against inadvertent double-counting (i.e. accounting for land use activities that have already been accounted for in some other way.) For land use activities, you need to pay special attention to the transportation planning and SIP land use assumptions to ensure that the reductions are above and beyond the existing land use assumptions.

Many factors can influence the accuracy and/or quality of your predictions of travel activity effects. Some factors influencing the confidence you have in your estimate include:

♦ Analysis zone size   The smaller the zone, the better able the model can predict interactions.

---

[30] For more information on the LUTRAQ study and model modifications, visit the 1000 Friends of Oregon web site at http://www.friends.org/resources/lut_reports.html .

[31] For more information on the MOBILE model visit EPA's Office of Transportation and Air Quality web address at:http://www.epa.gov/otaq/models.htm.. For more information on the EMFAC model, visit the California Air Resource Board's web address at: http://www.arb.ca.gov/msei/msei.htm.

- Socioeconomic and travel data quality

  The more detailed your data, the more accurately you can characterize the impacts of your policy/project.

- The age of your data.

  The older the data, the less confidence you should have in your results.

- Mid-course corrections

  If there is a mechanism for checking on the progress toward completion of land use activities, you may have better confidence the reductions will occur.

- Stated enforcement actions

  If there are discreet enforcement actions or remedies to address the likelihood of a project actually occurring, this will impact your confidence.

- Contingency remedy plans

  If you have a plan to remedy any shortfalls that may occur, this will increase the confidence you have in your emission estimates.

EPA recommends that you consider the confidence you have in your estimates when accounting for a land use activity in the SIP, and that you use reasonably conservative estimates of the benefits unless there is a comprehensive program including mid-course corrections, enforcement, and contingency remedies.

The VMEP policy was crafted in part to allow new and innovative programs with less established quantification methodologies to be included in a SIP. Most initial attempts to include land use activities as control strategies in SIPs will fall into this category.

The VMEP policy does not eliminate the need to use the most advanced quantification techniques where they exist. If you can use existing land use models and travel demand models, i.e. the four-step process, you should do so. For measures that do not lend themselves to quantification by the four-step process you can use other reasonable methodologies to quantify the emission benefits.

There may be cases where a land use activity has air quality benefits, but the scale of the activity is too small to be picked up by a travel demand model. For example, a local government may adopt a policy requiring microscale design elements to make neighborhoods more bike- and pedestrian-friendly. While this policy may have air quality benefits (because more people could choose to walk or bike instead of driving a car to make short trips), the impacts of this policy are too small to capture in the travel analysis zones analyzed in the travel demand modeling process. In cases where it is not possible to model the effects of land use policies and projects at a regional level, the emission reductions could be quantified in your SIP using an off-model technique. Forecasting procedures have been developed by some urban areas to account for travel demand changes based on micro-scale design.[32] However, developing or adopting such procedures may not be feasible for all urban areas. Off-model analyses could be used to estimate the travel and emissions impacts of micro-scale design, but should be evaluated and agreed upon through consultation with your EPA regional office.

---

[32] For more information, refer to the DOT draft report, "Data Collection and Modeling Requirements for Assessing Transportation Impacts of Micro-Scale Design," prepared by Parson Brinkerhoff Quade & Douglas, December 1999 (DTFH61-95-C-00168).

7.8 TRANSPORTATION CONTROL MEASURES AS CONTROL STRATEGIES

7.8.1 What are Transportation Control Measures?

Transportation control measures (TCMs) are measures that are specifically identified and committed to in a State Implementation Plan(SIP). A list of sixteen TCMs is provided in Section 108(f)(1)(A) of the Clean Air Act. In addition, any other measures adopted for the purpose of reducing emissions or concentration of air pollutants from transportation sources by reducing vehicle use, changing traffic flow, or mitigating congestion conditions may be considered TCMs. TCMs may be voluntary programs, incentives, regulatory programs, and/or market based/pricing programs. Note that, for the purposes of conformity, vehicle technology-based or fuel-based measures are not TCMs.

TCMs may be included in a SIP to demonstrate attainment of the NAAQS. In areas where TCMs are included in the SIP, the state or the MPO must make sure that all TCMs are funded in a manner consistent with the SIP's implementation schedule. The transportation conformity process is designed to ensure timely implementation of TCMs, thus reinforcing the link between SIPs and the transportation planning process. If the implementation of a TCM is delayed or if the TCM is only partially implemented, areas are required to make up the shortfall by either substituting a new TCM or other control measures.

7.8.2 How are Transportation Control Measures and land use activities related?

A land use project, such as a site-specific land use development, generally should not be considered a TCM. However, transportation aspects of specific land use projects such as transit stations or bike facilities can be considered TCMs.

EXAMPLES OF TCMS THAT SUPPORT LAND USE ACTIVITIES:

- Parking management programs
- Area-wide ride-share incentives
- Improved public transit
- Bicycle and pedestrian measures
- Park-and-ride programs

By designating transportation components of a land use activity as TCMs, the local area provides additional evidence that the overall land use activity will actually occur. TCMs are supportive and complementary to land use activities.

Transportation projects or TCMs that support land use projects may be eligible for funding under the Congestion Mitigation and Air Quality Improvement (CMAQ) program if they meet the funding eligibility criteria listed in the CMAQ program guidance[33]. The CMAQ program was created to provide funding for

---

[33] FHWA/FTA, The Congestion Mitigation and Air Quality Improvement (CMAQ) Program under the Transportation Equity Act for the 21st Century (TEA-21) Program guidance, April 28, 1999.

transportation projects that reduce emissions in nonattainment and maintenance areas. Congestion mitigation is another goal of the CMAQ program. Congestion relief can contribute to air quality improvements by reducing travel delays, engine idle time, and unproductive fuel consumption. CMAQ funds have been used for transportation projects that improve traffic flow, projects that enhance transit, and projects that encourage alternatives to driving alone. The Transportation Equity Act for the 21st Century has authorized $8.1 billion for the CMAQ program from 1998 to 2003. Transportation projects that are included in the SIP as TCMs and meet the CMAQ criteria are eligible for CMAQ funding.

Additional information on TCMs can be found in the following two EPA documents:

- "Transportation Control Measure: State Implementation Plan Guidance," EPA 450/2-89-020, September, 1990; and

- "Transportation Control Measure Information Documents," EPA 400-R-92-006, March 1992.

CHAPTER 8 INCLUDING LAND USE POLICIES OR PROJECTS IN THE CONFORMITY DETERMINATION WITHOUT HAVING THEM IN A SIP

8.1 WHAT IS A CONFORMITY DETERMINATION?

A conformity determination is a finding made by the metropolitan planning organization (MPO) or the state department of transportation and then subsequently by the U.S. DOT (FHWA/FTA) on the transportation plan, TIP, and projects in nonattainment and maintenance areas. The purpose of a conformity determination is to ensure that future transportation activities will not:

- Create a new air quality violation;
- Increase the frequency or severity of an existing air quality violation; or
- Delay timely attainment.

Transportation plans, TIPs, and projects in nonattainment and maintenance areas that are funded or approved by the FHWA and FTA must be found in conformity with the SIP in accordance with the requirements of the transportation conformity rule (40 CFR parts 51 and 93). (See section 3.5 for an explanation of plans and TIPs.)

8.2 HOW IS CONFORMITY DEMONSTRATED?

Conformity on plans, TIPs, and projects is demonstrated when the criteria and procedures established in the transportation conformity rule are satisfied. The transportation conformity rule requires a regional emissions analysis be conducted for all non-exempt projects included in the transportation plan and TIP. In the regional emissions analysis, the emissions from future transportation activities are estimated or modeled, just as they are when creating or revising a SIP's motor vehicle emission budget(s). These estimated emissions are compared to one of the following:

- If an area has a SIP that establishes a motor vehicle emissions budget(s), the estimated emissions produced by transportation activities must be shown to be less than or equal to the budget(s).

- When budgets aren't available, the estimated emissions are compared to either emissions from the "no-build" scenario, and/or emissions from a prior year (the specific requirements depend on the pollutant and the area's classification).

In CO and PM-10 nonattainment and maintenance areas, project level hot-spot analysis of localized air quality impacts are required before the project can be funded or approved by FHWA and FTA.

8.3 DOES THIS GUIDANCE IMPOSE NEW REQUIREMENTS FOR INCLUDING LAND USE ACTIVITIES IN A CONFORMITY DETERMINATION?

No, there are no new conformity requirements created by this guidance. The intent of this chapter is to generally capture how land use activities are currently being included within conformity determinations. Areas should use this guidance as a reference as new land use activities are introduced and existing land use activities are being implemented. The interagency consultation process should be used to ensure that this guidance is followed for new conformity determinations.

8.4 IF I HAVE INCLUDED A LAND USE ACTIVITY IN A SIP, DOES IT HAVE TO BE INCLUDED IN THE CONFORMITY DETERMINATION?

Yes. Any land use activity that was included in the SIP with associated air quality benefits should also be accounted for in subsequent conformity determinations, to the extent that it is being implemented according to the schedule in the SIP or still scheduled to occur.

8.5 CAN I ACCOUNT FOR THE EMISSIONS BENEFITS OF LAND USE ACTIVITIES IN A CONFORMITY DETERMINATION WITHOUT HAVING THEM IN A SIP?

Yes. Land use activities do not have to be included in a SIP. You can account for the emission reductions of a land use activity in a conformity determination, without having included it in any way in a SIP (see section 93.122(b)(1) of the transportation conformity rule). Section 8.16 of this chapter discusses the advantages of doing so.

8.6 HOW ARE LAND USE ACTIVITIES INCLUDED IN THE CONFORMITY DETERMINATION?

Note that this section, as well as sections 8.7 and 8.8, applies to areas that use network-based travel models for their conformity determinations. See section 8.15 if your area does not use a network model.

Land use activities can be included in a conformity determination either as **land use assumptions** or **control strategies**, depending on the case. Both land use assumptions and land use control strategies can affect the location of population and employment; their effects on population and employment should be integrated together before running the transportation model for the regional analysis.[34]

- **Land use assumptions:** The regional emissions analysis includes land use assumptions. These land use assumptions are made in the same way as those in the initial forecast of the SIP, discussed in chapter 6. Land use assumptions have to be reasonable, based on the best available information, and be consistent with the planned transportation system, pursuant to sections 93.110 and 93.122 of the conformity rule.

- **Control strategies:** The regional emissions analysis also includes the effects of adopted "control strategies." Control strategies are specific strategies for reducing emissions. Control strategies that are included in the conformity determination must meet certain requirements, discussed below.

---

[34] The conformity rule states that serious, severe, and extreme ozone nonattainment areas and serious CO nonattainment areas with an urbanized area population over 200,000 must use a travel demand model for their regional emissions analysis. In addition, any area already using a travel demand model must also use it for conformity. Areas without network-based travel models use other appropriate methods for estimating VMT.

Regardless of whether land use activities are considered land use assumptions or control strategies, there needs to be some type of assurance that they will occur before you include them in the conformity determination, and you can only include them to the extent that they are being implemented. The type of assurance that is necessary is discussed in the rest of this chapter.

8.7 WHAT ARE THE TRANSPORTATION CONFORMITY RULE'S REQUIREMENTS FOR LAND USE ASSUMPTIONS?

Some of the land use activities highlighted in this guidance could fall into the category of land use assumptions. Land use assumptions are the assumptions about where future population and employment will be located within a region. According to the conformity rule, assumptions must be:

Reasonable: Areas have to make reasonable assumptions regarding the distribution of employment and residences in the area (40 CFR 93.122(b)(1)(iii)). EPA and DOT believe that historical trends and recent data should be considered primary sources of information from which land use assumptions should be based and evaluated.

ILLUSTRATION: IS THERE A REASONABLE EXPLANATION FOR THE ASSUMED LAND USE CHANGE?

- In Chicago, land use forecasting is done by the Northeastern Illinois Planning Commission (NIPC), who give forecasts to the Chicago MPO and air quality planning agency for the State of Illinois. Chicago's most recent SIP and transportation plan conformity determination included assumptions that "the past trends of decentralized land use would be moderated" -- that is, there would be increased infill in the central part of Chicago. NIPC made these assumptions based on their judgement that the actions already underway and actions likely to be implemented would contribute to substantial reinvestment in existing communities and increased redevelopment would continue to occur. Though these assumptions were somewhat different from previous assumptions, NIPC provided adequate explanation and documentation for the change. In addition, the current land use plan generally supported this type of development and a substantial amount of infill development was already underway. Both EPA and DOT believed the assumptions to be reasonable, so they were included in the regional emissions analysis for the conformity determination.

- (Hypothetical example) The local governments of an area are currently discussing whether they want to establish an urban growth boundary. Many of the local governments are willing to adopt it for a variety of reasons, such as saving farmland and natural areas. However, some of the local governments are opposed because they do not want to limit additional growth. The MPO includes the boundary in the area's conformity analysis with a commitment to its implementation in the documentation for the conformity determination. However, the MPO's commitment isn't sufficient for the assumption to be considered reasonable, because ultimately the MPO does not have authority over land use and cannot implement the boundary. The urban growth boundary hasn't been adopted by all of the local governments; therefore, it cannot be included as a complete boundary in the conformity determination. It could only be applied in the specific geographic areas that adopted it.

"Best and latest available:" Areas need to use the best, most up to date information they have about future land use assumptions. The conformity rule states "land use, population, employment, and other network-based travel model assumptions must be based on the best available information" (40 CFR 93.122(b)(1)(ii)). Conformity determinations "must be based upon the most recent planning assumptions in force at the time of the conformity determination" (40 CFR 93.110(a)). Estimates of current and future population and employment are developed by the MPO or other agencies authorized to make such estimates, and approved by the MPO (40 CFR 93.110(b)).

ILLUSTRATION: ARE THE ASSUMPTIONS THE BEST AVAILABLE?

- A rapidly growing area has had a population growth rate of between 2.5 and 4% per year over the last ten years, and a corresponding increase in the number of jobs. The urbanized area has increased 80% over this same period. The MPO assumes that land will be consumed more slowly in the future, and forecasts that the land consumption rate for the next ten years will only be half of what it was, reasoning that the current building boom won't last forever.

    This change in future land consumption rate would not be the best available assumption. Unless there were some compelling evidence for assuming that land consumption will drop (e.g., the area has adopted an urban growth boundary), the best available assumptions would be based on the most recent trends. In the situation described here, there is insufficient evidence to support an assumption that the current trends won't continue.

Consistent with planned transportation system: The conformity rule also states that scenarios of land development and use must be consistent with the future transportation system planned. The distribution of employment and residences throughout the area must be reasonable given the transportation network planned (40 CFR 93.122(b)(1)(iii)).

ILLUSTRATION: IS THE FORECASTED LAND USE CONSISTENT WITH PLANNED TRANSPORTATION?

- An area plans to build a new highway beltway. They forecast additional population and employment to locate around the beltway after it is completed. These assumptions are consistent with the transportation system planned.

- An area plans to build a new transit line with a series of new transit stops. They forecast increased population and employment around the transit stops. These assumptions would be consistent with the new transportation project planned, particularly if other actions, such as policies to facilitate transit-oriented development, are adopted to encourage development around transit.

- In the example above, the transit stops will not be completed for 10 years, but the MPO forecasts increased population and employment around the transit stops in five years. These assumptions could not be used because they are inconsistent with the planned transportation system, unless there were other adopted policies to encourage development in these areas before the transit stops are built.

8.8 How are the land use assumptions in a conformity determination reviewed?

The interagency consultation process should be used to evaluate and choose the assumptions to be used in the regional emissions analysis for conformity.[35] Regardless of whether land use modeling or best judgment of planners is used to arrive at what future land use will be, the **interagency consultation partners should agree that the assumptions are reasonable, best available, and consistent with the transportation system planned**. See the above examples for determining appropriateness of assumptions.

As stated previously, land use assumptions have to be based on the latest and best available information. Keeping this requirement in mind, we would expect that land use assumptions made for a conformity determination would be generally consistent with the trends assumed in the previous conformity determination or those included in a recently submitted SIP. This expectation is a result of the fact that land use trends can change slowly. If the trends are similar to those from the previous conformity determination or a recently submitted SIP, no additional assurance about assumptions is probably necessary. The fact that the trends are similar is, in effect, assurance that the assumptions made are reasonable, and likely to occur.

However, if land use assumptions are radically different from historical trends reflected in previous assumptions, the consultation process should be used to determine why these assumptions are appropriate. The conformity determination would have to document and explain why the assumptions are appropriate. The documentation should be made available for public comment during the conformity determination process. If the conformity documentation doesn't provide a reasonable explanation, then the conformity determination will be closely scrutinized, and may not be approved.

In subsequent conformity determinations, land use assumptions should be reevaluated through the interagency consultation process. If a conformity determination's land use assumptions differ significantly from past trends, the interagency consultation parties should pay close attention to land use assumptions made in subsequent conformity determinations. Assumptions can only continue to be used to the extent they are being implemented or still on schedule as planned.

8.9 What are control strategies?

A control strategy is a project, program, or activity undertaken for the purpose of reducing the amount or the concentration of emissions. For example, some cities use reformulated gasoline as a strategy for controlling motor vehicle emissions. Other examples of control strategies are retrofitting heavy duty diesel trucks to produce less emissions, increased provision of transit, and commuter choice programs. Land use activities can also be control strategies. (The term "control strategies" is not synonymous with the term "transportation control measures." See section 7.8 for more about transportation control measures.)

---

[35] Interagency consultation is required by the conformity rule (40 CFR 93.105). For more information on interagency consultation, visit the FHWA document, "Transportation Conformity: A Basic Guide for State and Local Officials" at http://www.fhwa.dot.gov/environment/conformity/basic_gd.htm.

8.10 WHAT ARE THE CONFORMITY RULE'S REQUIREMENTS FOR CONTROL STRATEGIES?

Basically, control strategies must be committed to by the appropriate jurisdiction before they can be included in the regional analysis for a conformity determination. In 40 CFR 93.122(a)(3) and (4), the rule states that:

- If the control strategy requires regulatory action to be implemented or undertaken, it can be included in a conformity determination if:
  - the regulatory action is already adopted by the enforcing jurisdiction;
  - the strategy has been included in an approved SIP; or
  - there is a written commitment to implement the strategy in the submitted SIP.

- If the control strategy is not included in the transportation plan and TIP or the SIP, and it does not need a regulatory action to be implemented, then it can be included in the conformity determination's regional emissions analysis if the conformity determination contains a written commitment to implement it from the appropriate entities.

As is the case with land use assumptions, the conformity analysis can only account for approved control strategies to the extent that they are being implemented.

8.11 HOW DO I DETERMINE WHETHER A LAND USE ACTIVITY IS A LAND USE ASSUMPTION OR A CONTROL STRATEGY?

We realize that it may be difficult to determine whether a land use activity is a land use assumption or a control strategy. In general, if a land use activity is adopted and implemented above and beyond what has already been included in the land use assumptions, and emissions benefits have been identified for the specific activities, it can be regarded as a control strategy. Another consideration that may help clarify whether a land use activity is a land use assumption or a control strategy is its purpose:

- Is the primary purpose of the land use activity to improve air quality? If so, it likely falls into the category of control strategy.

- Is the primary purpose of the land use activity to reduce emissions for conformity analyses? If so, it likely falls into the category of control strategy.

These questions are only intended to be guidelines. You should discuss the decision with the other participants in the interagency consultation process if you have doubt about which category fits a particular project or policy best.

Regardless of whether you call a land use activity an assumption or a control strategy, it has to be based in reality – if your land use forecast differs significantly from the past trends, there must be adequate justification for the change.

8.12 WHAT ARE SOME EXAMPLES OF LAND USE ACTIVITIES THAT FIT IN EACH CATEGORY?

It is not always easy to determine into which category a land use activity would fit. Either category could be appropriate, depending on the circumstances. Below are some examples to illustrate this point.

Examples of Land Use Assumptions:

Urban Growth Boundary:
- In recent years, the local governments that make up the Denver region have agreed to an urban growth boundary. Approximately 85% of the local governments have signed formal agreements to adhere to this boundary. The others have verbally agreed to comply. In this case, the combination of written and verbal agreements satisfies the requirement that the urban growth boundary is a "best available" land use assumption. Although this assumption was new, there was sufficient evidence to document that all of the local governments are implementing the boundary, and therefore we consider it an appropriate assumption to make. The consultation process will be used to review the implementation of the boundary for future conformity determinations.

Transit Oriented Development (Hypothetical example)
- An area decides to accommodate future growth along a particular corridor, currently agricultural land, and they include funding to build a light rail line and stations in their transportation plan and TIP. Through the consultation process, the area decides to concentrate higher density development around these stations. They include an explanation and appropriate documentation in the conformity determination that the local governments have agreed to the approach and have committed to a schedule for changing their zoning to make it occur on the timeline assumed in the conformity analysis. The explanation is supported with details from the local governments' economic growth and incentive plans. In this hypothetical example, the transit oriented development could be a land use planning assumption: it is based on reasonable information and the land use scenario is consistent with the planned transportation infrastructure. Because the plan to focus development was discussed and agreed to through the consultation process and documented in the conformity determination, the assumption could be included in the emissions analysis for conformity.

Examples of Land Use Control Strategies:

Urban Growth Boundary:
- In 1973, the State of Oregon passed a planning statute that requires local governments to establish an urban growth boundary. Because of this law and its implementation, Portland's MPO can include the urban growth boundary as a control strategy in the emissions analysis done for their SIPs and conformity determinations, because the statute is in place and is being implemented.

Parking Requirements
- An area decides they want to set a maximum on the amount of parking that can be built for new residential or commercial development. Before the effects of the parking requirement could be included in an emissions analysis for a conformity determination, it would have to be adopted by the jurisdiction that has the power to enforce it.

Transit Oriented Development (Hypothetical examples):
- (1) An area decides to accommodate future growth along a particular corridor, currently a low density commercial one, and the transportation plan and TIP includes the funding to build a light rail line and stations along this corridor. However, the local governments have not yet taken any actions to implement transit-oriented development along this corridor.

- (2) An area decides to accommodate future growth along a particular corridor. Currently, a light rail line already exists in this corridor but because there is low density development surrounding it, the light rail line is underutilized. However, the local governments have not yet taken any actions to implement transit-oriented development along this corridor.

In these two hypothetical examples, rather than converting undeveloped land to high density development, the area would be redeveloping an existing corridor. In these cases, we may not consider transit-oriented development to be a planning assumption. A greater amount of political will would be needed for the planned changes to take place, and therefore we would want a greater degree of commitment to ensure that the development occurs. In these types of cases, EPA would regard transit-oriented development as a control strategy that would need to be adopted by the enforcing jurisdictions -- the local governments -- before it could be included in an emissions analysis for a conformity determination.

8.13 WHAT IS "DOUBLE COUNTING?"

EPA wants to ensure that areas do not count the effects of a land use activity twice. Areas must be sure that what they are including in the conformity determination has not already been included in some other way. A particular land use activity could be included either as an assumption or as a control strategy, but not as both an assumption and as a control strategy since that would be counting it twice. Similarly, an area should include either the effects of a land use policy, or the effects of the individual projects that happen as a result of that policy. It should not count both the policy and its resulting projects since that would be counting the effects twice.

For example, suppose a metropolitan region adopts a policy to give incentives to developers for building infill development in downtown. The area can then include the likely results of that policy into the land use assumptions for the conformity determination, such as increased population and employment in the zones that would be affected by the policy. Once that is done, however, it would not be appropriate to add new population and employment for the individual developments that occur as a result of that policy. That would be double counting, because the new population and employment that result from the individual projects have already been accounted for in the conformity determination when the policy was included.

Likewise, if instead you have already included the effects of an enormous new development into the conformity determination, it would not be appropriate to also include the effects of the policy that caused the specific development to occur. Either the effects of one or the other should be counted, but not both.

8.14 WHAT IF A LAND USE ACTIVITY IS TOO SMALL TO HAVE AN IMPACT ON THE OUTCOME OF TRAVEL DEMAND MODELING?

There may be some land use activities that have an air quality benefit, but their effects are too small to be picked up by a travel demand model. In cases where it is not possible to model the effects of land use policies and projects at a regional level, the emissions reductions could be quantified in your conformity determination using an off-model technique. Forecasting procedures have been developed by some urban areas to account for travel demand changes based on micro-scale design[36]. However, developing or adopting such procedures may not be feasible for all urban areas. Off-model analyses could be used to estimate the travel and emissions impacts of micro-scale design, but should be evaluated and agreed upon through interagency consultation of the MPO, state and local air quality planning agencies, state and local transportation agencies, EPA, and DOT.

8.15 WHAT IF OUR AREA DOESN'T USE A TRAVEL DEMAND MODEL FOR TRANSPORTATION PLANNING?

There are some areas that are not required to use travel demand forecasting models. In these areas, the emission reductions associated with land use activities could be quantified in your conformity determination using another technique, consistent with 40 CFR 93.122(c), and be chosen through the interagency consultation process. However, land use assumptions must still be reasonable, based on the best available information, and consistent with planned transportation. Land use control strategies must meet the requirements outlined above.

8.16 WHAT ARE THE ADVANTAGES OF ACCOUNTING FOR LAND USE ACTIVITIES IN THE CONFORMITY DETERMINATION WITHOUT HAVING THEM IN THE SIP?

First, conformity determinations offer more opportunities to account for land use activities as they happen. Conformity must be redetermined at least every three years. In contrast, SIPs are generally prepared at a single time. (Revisions can be made to a SIP at a later date, and you may be required to monitor and evaluate programs and make corrections.)

Second, a conformity determination looks at the effects of the land use and transportation system many more years into the future, because it must examine the life of the transportation plan.[37] This is in contrast to SIPs: attainment demonstrations only look as far as the attainment date, which is at most 7 years in the future; maintenance plans require maintenance of the standards for two consecutive time periods of 10 years each. It may take more than 10 years for land use policies or projects to have an impact on travel decisions and therefore air quality; the conformity determination looks at a time frame in which you can see their effects.

Third, an MPO might prefer to have effects of land use activities in a conformity determination that haven't

---

[36] For more information, refer to the DOT draft report, "Data Collection and Modeling Requirements for Assessing Transportation Impacts of Micro-Scale Design," prepared by Parson Brinkerhoff Quade & Douglas, December 1999 (DTFH61-95-C-00168).

[37] DOT's metropolitan planning regulations require plans to have at least a 20 year planning horizon. Some areas adopt transportation plans that cover more than 20 years. The plans must be updated every three years.

been accounted for in the SIP. These reductions are then "surplus" to the SIP and could be used to offset the emission-creating effects of other projects in the transportation plan.

Finally, another advantage of including land use activities in conformity rather than in a SIP is the ease of accommodating changes in the land use activity. If the features of the land use activity produce fewer emissions than originally expected, or if the activity becomes delayed, the change would simply need to be reflected in the next conformity determination. You wouldn't have the problem of having to make up a SIP "shortfall"-- that is, you would not have to revisit your SIP to make up the emissions reductions. However, you would have to revisit and revise your transportation plan and TIP and make up the reductions from these programs unless other agreements are reached with the state air agency. You would also need to be sure that the activity is correctly reflected in the next conformity determination.

CHAPTER 9 ADDITIONAL CONSIDERATIONS WHEN ACCOUNTING FOR LAND USE ACTIVITIES IN THE SIP OR THE CONFORMITY PROCESS

9.1 HOW CAN I DETERMINE WHETHER OR NOT MY LAND USE ACTIVITIES MIGHT HAVE AIR QUALITY BENEFITS?

Sketch planning techniques can be very useful in helping you identify the land use activities you want to implement. Sketch planning tools may allow you to determine the relative magnitude of emission reductions. These techniques are likely to require fewer resources and less time than a SIP quality estimate for each policy or project. This level of quantification can be particularly useful if you are comparing two or more policies or projects to determine the appropriate one(s) to implement.

EPA's SMART GROWTH INDEX

EPA's Smart Growth Index (SGI) is a sketch model that evaluates transportation and land use alternatives and assesses their impact on travel demand, land consumption, housing and employment density, and pollution emissions using geographic information system (GIS) technology. SGI generates predictions that can help localities understand the environmental implications of different development plans. It is intended as a planning support tool.

In particular, SGI has been developed in response to expressions of need for simplified scenario-testing tools that can be applied rapidly and inexpensively in areas that do not have access to sophisticated land use or transportation planning models such as ITLUP or TRANUS, or that wish to perform quick sketches prior to applying such models.

As a sketch tool, SGI has limitations that need to be clearly understood by prospective users. It is not:

- A highly technical model, such as TRANUS or ITLUP, that attempts to simulate integrated land use/transportation dynamics with a high degree of mathematical precision, especially for regulatory compliance or major investment evaluation purposes.

- A land economics model that considers land price effects on growth patterns.

- A calibrated transportation planning model suitable for evaluating major transportation system improvement alternatives.

- A traffic engineering or highway design model.

SGI does borrow or adapt certain elements and methods from these and other place-making tools, but its limitations need to be recognized. Ideally, the use of SGI as a preliminary evaluation tool will lead to justification and resources for advanced analysis with the types of tools described above.

However, the results of sketch planning are generally not rigorous enough for use as SIP or conformity quantification tools.

You do not necessarily need to perform "SIP quality" quantification for the quantification to be useful. You may determine for some reason that you do not want or need to explicitly include a land use activity as a control strategy in the air quality planning process, or simply, you may be trying something new, which you cannot quantify with relative certainty.

If you determine that a land use activity is directionally beneficial, you may decide it is worth implementing. You may find that you are able to more accurately quantify the benefits of the land use activity after you have had some experience implementing it. Many areas have included control strategies in their SIP, without explicitly accounting for emissions benefits from them, as measures that will assist in meeting or maintaining air quality standards.

For more information on available sketch modeling tools that can be used at the state and local levels, and an analysis of the ease or difficulty of their use, see EPA600/R-00/098, "Projecting Land Use Change: A Summary of Models for Assessing the Effects of Community Growth and Change on Land use Patterns."

9.2 HOW WILL THE TIME FRAME FOR IMPLEMENTING THE LAND USE ACTIVITIES AFFECT WHICH ACCOUNTING OPTION I CHOOSE?

Areas that are designated nonattainment have a defined period of time to reach their air quality goals. Depending on the land use activity you are implementing, you may find that the benefits won't occur until after you are required to attain.

For example, in a region that has been experiencing considerable sprawl-type development over many years, land use policy actions to encourage high density, mixed use development in existing urban cores may take ten to 20 years to have an significant impact on development trends and emissions from motor vehicle travel. Therefore, if your attainment date was less than ten years from now, it may not be beneficial to include these land use activities as control strategies in your attainment SIP. However, you may wish to include these land use activities in the SIP without quantifying emissions reductions for them. Many areas have included control strategies in their SIP, without explicitly accounting for emissions benefits from them, as measures that will assist in meeting or maintaining air quality standards. To do so may allow you to highlight your land use activities and help ensure that the control strategies are supported by all levels of government.

In cases where land use activities are not expected to achieve emissions reductions within the attainment SIP's time frame, but are expected to achieve emissions reductions in the future, these land use activities could be included in a future maintenance plan SIP. The maintenance period is 20 years, covered by two ten-year maintenance plans (the second is submitted eight years after the first). Alternatively, you may choose to take account of the land use activity in a conformity determination, which involves a planning horizon of at least 20 years.

While many policy decisions take long periods of time to yield emission reductions, specific projects initiated by private developers and financial institutions may occur more rapidly. For example, numerous developers are building housing communities, office complexes, and shopping centers which build in principles such as high density, orientation near public transit, reducing the number of parking spaces, and designing streets and buildings to encourage walking. These specific projects may yield localized emission benefits in a shorter time frame than regional policies. Projects that are large scale or that are constructed in phases would take a longer time to show benefits.

A combination of short-term localized projects and longer-term, regional policy, incentive, and education strategies is the best approach to creating significant benefits.

9.3 WHAT OTHER IMPORTANT ISSUES SHOULD I BE AWARE OF IN QUANTIFYING AIR QUALITY BENEFITS?

There are a number of additional issues you should consider when quantifying the air quality benefits of land use policies and projects. These include:

- The synergistic or antagonistic interactions between policies and projects;
- Deciding whether to quantify the benefits of policies and projects individually, or with other policies and projects;
- Ensuring that your estimates are conservative enough, and the potential consequences of not being conservative; and
- The effect of the scale of your project on the uncertainty associated with your estimate of emissions reductions.

9.3.1 Accounting for interactions between land use activities

Relationships between land use and travel are complex. Land use activities may interact with each other, either enhancing the emissions reductions they achieve, or in some cases diminishing the beneficial effects.

For example, microscale changes such as adding sidewalks and bike paths may not be enough by themselves to alter vehicle ownership or mode choice if the region is largely vehicle-oriented. A combination of actions, such as mixed-use development, increased access to transit, and pedestrian improvements, may yield more success in shifting travel activity from vehicles to other modes.

In addition, in areas where congestion is a current problem, increased density of development may yield increases in localized emissions.

It is important to consider the potential interactions of certain land use activities with other control strategies, or with current conditions when you are quantifying the emission effects.

9.3.2 Quantifying land use activities individually or as a group

Calculating the emission benefits of land use control strategies together can result in fewer modeling steps. If you model several strategies together, you will need to calculate an initial forecast of future emissions without the land use activities, and compare that to a calculation of the emissions with all of the land use activities included. Quantifying the effects of land use activities together will likely better capture the interactions and synergies produced by several land use policies or projects.

If you model land use activities individually, you will need to run a separate calculation for each activity. In order to take into account the interactive effects described above, whenever feasible, you should run the model accounting for previously modeled land use activities in your baseline. That is, you should estimate the benefits of land use activities in a series, continuously building on previous land use activities.

By determining the explicit amount of emissions reduced by a single land use activity, you can determine the air quality benefits of that control strategy and the cost effectiveness of pursuing it. This will allow you to compare several land use control strategy options in terms of emission reductions and cost effectiveness. In addition, by going through the process of calculating the benefits of the control strategy, you are likely to gain more insight into how the land use control strategy is working. Calculating a specific benefit for the control strategy can also help explain the effectiveness of the control strategy to the public. This can be particularly important if quantifying the control strategy's effectiveness is needed for it to be adopted.

9.3.3 Using conservative estimates

Modeling the air quality impacts of land use activities is an inherently uncertain process, and it is important that you are confident in the emissions benefits that you account for in a SIP or conformity determination. If you overestimate the emission reductions for land use activities, you ultimately may not meet the air quality standards. If a land use activity does not result in the expected emission benefits, you will have to find new reductions to make up the shortfall.

9.3.4 Taking into account the scale of the land use activity

Land use and transportation models are likely to more accurately predict the benefits of a land use project or land use policy when its impact is large in scale. For larger-scale projects, the law of averages will tend to even out outlying discrepancies, whereas smaller-scale projects will be more affected by individual variation. You need to be aware that for very small projects and policies, the inherent errors in the modeling can even be greater than the modeled emission benefits of the project. The scale of the project can therefore greatly impact the confidence levels of the modeling, and thus the amount of emissions reductions you claim.

9.4 HOW WILL EPA ASSIST ME WITH QUANTIFICATION?

Staff in your EPA regional office will work with you to ensure that the quantification of your land use activities will meet all the applicable statutory and regulatory requirements for inclusion in the SIP or conformity determination.

Beyond the general guidelines contained within this guidance document, EPA is developing additional guidance documents outlining quantification methodologies for individual types of land use activities to assist you in addressing quantification issues specific to those policies and projects. The first of these guidance documents, "Comparing Methodologies to Assess Transportation and Air Quality Impacts of Brownfields and Infill Development" (EPA 231-R-01-001) focuses on quantifying the benefits of infill development and brownfield redevelopment for SIP purposes. EPA will continue to develop quantification methodologies, and will release other policy-specific guidance documents over time, and these materials will be made available via the following web address: http://www.epa.gov/oms/transp/traqsusd.htm

In addition, you may find the following resources useful:

- U.S. EPA, 1997. Evaluation of Modeling Tools for Assessing Land Use Policies and Strategies, EPA 420-R-97-007. U.S. Environmental Protection Agency, Office of Mobile Sources, Ann Arbor, MI.

- U.S. EPA, 2000. Projecting Land-Use Change: A Summary of Models for Assessing the Effects of Community Growth and Change on Land-Use Patterns. EPA 600-R-00-098. U.S. Environmental Protection Agency, Office of Research and Development, Cincinnati, OH.

- U.S. DOT, 2000. Data Collection and Modeling Requirements for Assessing Transportation Impacts of Microscale Design. Parsons, Brinkerhoff, Quade & Douglas, Inc. for US Department of Transportation, Washington, DC.

- National Cooperative Highway Research Program, 1999. NCHRP 423 Land Use Impacts of Transportation: A Guidebook. National Academy Press.

- Transit Cooperative Research Program, 1999. Report 48--TCRP Web Document 9: Integrated Urban Models for Simulation of Transit and Land-Use Policies: Final Report by Eric J. Miller, David S. Kriger, and John Douglas Hunt; University of Toronto Joint Program in Transportation and DELCAN Corporation, Sponsored by the Federal Transit Administration (FTA). National Academy Press.

- U.S. DOT, February 1995. Travel Model Improvement Program - Urban Design, Telecommuting, and Travel Forecasting- Conference Proceedings, Final report DOT-T-96-09, Final Report.

- U.S. DOT, July 1998. Travel Model Improvement Program - Land Use Compendium. DOT-T-99-03. Engelke, L. Texas Transportation Institute.

- U.S. DOT, June 2000. Travel Model Improvement Program - Land Use Forecasting Studies. Parsons, Brinkerhoff, Quade & Douglas, Inc. for U.S. DOT, Washington D.C.

- U.S. EPA, 2000. Air Quality Impacts of regional Land Use Policies. Robert A. Johnson, University of California; John E. Abraham, University of Calgary, for the US Environmental Protection Agency.

SECTION 3: APPENDICES

Section 3 provides a variety of additional resources to support users of this guidance.

Appendix A	Examples of Land Use Policies and Strategies	A-1
Appendix B	Related Internet Web Sites	B-1
Appendix C	Related Work Efforts	C-1
Appendix D	Glossary of Terms	C-1
Appendix E	List of Acronyms	D-1
Appendix F	References to Relevant Policies, Guidance Documents, and General Information Sources	E-1
Appendix G	Regional and State Contacts	F-1

APPENDIX A   EXAMPLES OF LAND USE POLICIES AND STRATEGIES

This list of examples of land use strategies and policies has been borrowed from a June, 1995 report by JHK & Associates for the California Air Resources Board entitled, "Transportation-Related Land Use Strategies to Minimize Mobile Source Emissions: An Indirect Source Research Study." This report is available on the U.S. Department of Energy's "Sustainable Developments website at http://www.sustainable.doe.gov/pdf/arb-report/arb-overview.htm

Some examples of Land Use Strategies include:

- Concentrated activity centers: Encourage pedestrian and transit travel by creating "nodes" of high density mixed development, that can be more easily linked by a transit network.

- Strong downtowns: Encourage pedestrian and transit travel by making the central business district a special kind of concentrated activity center, that can be the focal point for a regional transit system.

- Mixed-use development: Encourage pedestrian and transit travel by locating a variety of compatible land uses within walking distance of each other.

- Infill and densification: Encourage pedestrian and transit travel by locating new development in already developed areas, so that activities are closer together.

- Increased density near transit stations: Encourage transit travel by increasing development density within walking distance (0.25 to 0.50 miles) of high capacity transit stations, and incorporate direct pedestrian access.

- Increased density near transit corridors: Encourage transit travel by increasing development density within walking distance (0.25 to 0.50 miles) of a high capacity transit corridor.

- Pedestrian and bicycle facilities: Encourage pedestrian and bicycle travel by increasing sidewalks, paths, crosswalks, protection from fast vehicular traffic, pedestrian-activated traffic signals, and shading.

- Interconnected street network: Encourage pedestrian and bicycle travel by providing more direct routes between locations. Also, alleviate traffic congestion by providing multiple routes between origins and destinations.

- Strategic parking facilities: Encourage non-automobile modes of transit by limiting the parking supply, and encourage carpooling by reserving parking close to buildings for carpools and vanpools.

Some examples of Land Use Polices include:

Encourage focused higher density by:

- Allowing transfer of unused development density capacity in outlying areas to permit development density above maximum limits near central areas and transit (zoning/regulations and non-monetary incentives);

- Allowing increased density for residential, retail, and employment generating uses in central areas and around transit (zoning/regulations and non-monetary incentives);

- Setting minimum densities for residential, retail, and employment generating uses in central areas and around transit (zoning/regulations);

- Requiring no net decrease in residential density for redevelopment (zoning/regulations);

- Stating densities in terms of square feet of land per dwelling unit, rather than minimum lot size, to encourage clustering (zoning/regulations);

- Granting incentives (e.g., reduced parking requirements, accelerated permit processing, infrastructure upgrades) for development that focuses on existing urban areas and infill (non-monetary incentives);

- Adjusting development impact fee structures or giving tax breaks to encourage infill and increased density development near transit and activity centers, and to discourage outlying development (monetary incentives).

Encourage mixed-use zones by:

- Allowing mixed use, which is now prohibited in many places (zoning/regulations);

- Requiring mixed uses, with certain percentages of residential, public, and commercial uses in target areas (zoning/regulations);

- Using fine-grained zoning to achieve mixed use while ensuring residential zones are buffered from heavy industrial zones with light industrial and commercial zones (zoning/regulations);

- Using mixed-use overlay zoning, to add a second use to an area that is primarily in another use, e.g., commercial corridors along major arterials in a primarily residential area (zoning/regulations);

- Granting incentives (e.g., reduced parking requirements, accelerated permit processing, infrastructure upgrades) for development that locates transit- or pedestrian-oriented amenities, like housing or child care near commercial uses and pedestrian-oriented design (non-monetary incentives);

- Adjusting development impact fee structures or giving tax breaks to encourage mixed use (monetary incentives).

Encourage pedestrian, bicycle, transit, and carpooling activity by:

- Requiring connected, narrower streets with trees and sidewalks in new development (zoning/regulations);

- Requiring bicycle lanes and transit stops on larger streets in new development (zoning/regulations);

- Requiring traffic-calming devices in new development; e.g., textured paving at crossings, frequent intersections with pedestrian-activated traffic signals, and traffic circles (zoning/regulations);

- Reducing requirements for setbacks and minimum lot sizes to create a stronger connection between buildings and sidewalks (zoning/regulations and non-monetary incentives);

- Requiring pedestrian scale signs in pedestrian- and transit-oriented areas (zoning/regulations);

- Reducing minimum parking requirements near transit hubs and for projects providing features that encourage pedestrian, bicycle, and transit activity (zoning/regulations and non-monetary incentives);

- Setting parking maximums in transit- and pedestrian-oriented areas (zoning/regulations);

- Requiring preferential parking for carpools (zoning/regulations).

For more examples of applications of land use activities that may reduce reliance on automobiles and thus the air quality impacts of driving, see the following sources:

Smart Growth Network's Case Studies page
http://www.smartgrowth.org/casestudies/casestudy_index.html

Sierra Club's "Smart Choices or Sprawling Growth: A 50 State Survey of Development"
http://www.sierraclub.org/sprawl/50statesurvey/intro.asp

Center of Excellence for Sustainable Development Land Use Planning Success Stories
http://www.sustainable.doe.gov/landuse/lusstoc.shtml

Urban Land Institute's Smart Growth: News, Tools and Hot Links
http://www.uli.org/indexJS.htm

Sprawl Watch Clearinghouse Best Practices page
http://www.sprawlwatch.org/bestpractices.html

White House Livable Communities web site
http://www.livablecommunities.gov

USEPA Office of Solid Waste and Emergency Response Brownfields website
http://www.epa.gov/brownfields

APPENDIX B   RELATED INTERNET WEB SITES

Local, Regional and State Government Organizations

The Council of State Governments, http://www.statesnews.org
> Founded on the premise that the states are the best sources of insight and innovation, CSG provides a network for identifying and sharing ideas with state leaders.

Environmental Council of the States, http://www.sso.org/ecos
> ECOS is a national non-profit, non-partisan association of state and territorial environmental commissioners.

International City/County Management Association, http://www.icma.org
> ICMA is a professional and educational association for more than 8,000 appointed administrators and assistant administrators serving cities, counties, other local governments, and regional entities around the world. Through the Smart Growth Network, it assists its members in identifying strategies and tools to protect the health and welfare of their communities through the integration of environmentally sound decision making and economic growth.

Local Government Commission, http://www.igc.org
> The LGC is a nonprofit membership organization that offers education, training, and technical assistance to local areas seeking to implement innovative long-term solutions that further economically and environmentally sustainable land use patterns.

National Association of Counties, http://www.naco.org
> NACo is a full-service organization that provides legislative, research, technical, and public affairs assistance to its members. NACo acts as a liaison with other levels of government, works to improve public understanding of counties, serves as a national advocate for counties and provides resources to counties.

National Association of Local Government Environmental Professionals, http://www.nalgep.org
> NALGEP is a nonprofit association representing local government officials who are responsible for ensuring environmental compliance and implementing environmental programs.

National Association of Regional Councils, http://www.narc.org/about.html
> Fostering regional cooperation and building regional communities, NARC is a nonprofit membership organization serving the interests of regional councils nationwide.

National Governors Association, http://www.nga.org
> NGA is a bipartisan national organization of, by, and for the nations' Governors. Through NGA, the Governors identify priority issues and deal collectively with issues of public policy and governance at both the national and state levels.

The State and Territorial Air Pollution Program Administrators/ Association of Local Air Pollution Control Officials, http://www.4cleanair.org/about.html
> STAPPA and ALAPCO are the two national associations representing air pollution control agencies in the U.S. The associations serve to encourage the exchange of information among air pollution control officials, to enhance communication and cooperation among federal, state and local regulatory agencies, and to promote good management of our air resources.

U.S. Conference of Mayors, http://www.usmayors.org/uscm

USCM is the official nonpartisan organization of cities with populations of 30,000 or more, collectively speaking on matters pertaining to organizational policies and goals.

Miscellaneous Local/State Government Sites

City of Austin, Smart Growth Initiative
http://www.ci.austin.tx.us/doorstep/98/10/smartgrow.htm#anchor1055467

New Jersey Pinelands Comprehensive Management Plan
http://www.state.nj.us/pinelands/cmp.htm

Oregon Transportation and Growth Management Program
http://www.lcd.state.or.us/issues/tgmweb/about/index.htm

Smart Growth in Maryland
http://www.op.state.md.us/smartgrowth

Federal Government Organizations and Initiatives

Clean Cities Program, http://www.livablecommunities.gov/toolsandresources/tr_clean_cities.htm
Sponsored by DOE, this program is a voluntary, locally-based government/industry partnership, that mobilizes local stakeholders in the effort to expand the use of alternatives to gasoline and diesel fuel by accelerating the deployment of alternative fuel vehicles (AFVs), and building a local AFV refueling infrastructure.

Congestion Mitigation and Air Quality Improvement Program,
http://www.livablecommunities.gov/toolsandresources/tr_cmaq.htm
Sponsored by FWHA, CMAQ funds projects in areas that do not meet the National Ambient Air Quality Standards (non-attainment areas) and former non-attainment areas that are now in compliance (maintenance areas) for ozone, carbon monoxide, and small particulate matter.

Department of Transportation, http://www.dot.gov/index.htm
DOT was established by Congress to serve the U.S. by ensuring a fast, safe, efficient, accessible and convenient transportation system that meets the nation's vital interests and enhances the quality of life of the American people, today and into the future.

DOT's Bureau of Transportation Statistics, http://www.bts.gov
BTS was established for data collection, analysis, and reporting and to ensure the most cost-effective use of transportation-monitoring resources.

EPA Community-Based Environmental Protection, http://www.epa.gov/ecocommunity
CBEP itegrates environmental management with human needs, considers long-term ecosystem health and highlights the positive correlations between economic prosperity and environmental well-being.

EPA Office of Air Quality Planning and Standards, http://www.epa.gov/oar/oaqps/cleanair.html
EPA's OAPQ directs national efforts to meet air quality goals and is responsible for implementing other major provisions of the Clean Air Act, including those related to visibility, permitting, and emission standards for a wide variety of industrial facilities.

Federal Highway Administration, http://www.fhwa.dot.gov/environment
The FHWA Planning and Environment Core Business Unit (HEP) provides policy direction and

guidance in three major areas including statewide and metropolitan transportation planning, human and natural environment, and real estate services.

FTA's National Transit Library, http://www.fta.dot.gov/ntl/index.html
>The Federal Transit Administration provides a collection of transit and transportation related articles developed by the FTA, DOT, and partners in the transit industry.

Livable Communities, http://www.livablecommunities.gov/toolsandresources/tr_lc.htm
>Sponsored by FTA, this program helps communities get involved in planning and designing transit systems that are customer-friendly, community-oriented and well designed.

Transportation Analysis and Simulation System in Washington, DC,
http://www.livablecommunities.gov/toolsandresources/tr_transims.htm
>Sponsored by DOT, EPA, and DOE, TRANSIMS is an advanced software package that would allow local planning agencies to simulate the movement of individuals and vehicles for an entire metropolitan region.

Nonprofit Organizations

American Farmland Trust, http://www.farmland.org
>AFT was founded to protect the nation's agricultural resources and does so by stopping the loss of productive farmland and by promoting farming practices that lead to a healthy environment.

American Planning Association, http://www.planning.org
>APA is a nonprofit, public interest organization representing 30,000 practicing planners, elected and appointed officials, and citizens involved in urban and rural planning issues.

Bicycle Federal of America, http://www.bikefed.org
>BFA is a national, nonprofit corporation working to create bicycle-friendly and walkable communities.

Brookings Institution Center on Urban and Metropolitan Policy, http://www.brook.edu/urban
>The Center has special expertise on regional governance issues and seeks to shape a new generation of urban policies that will help build strong cities and metropolitan regions. In collaboration with leading scholars and practitioners nationwide, the Brookings Center is launching a series of original research and policy projects to help inform national debates and provide practical policy options for key decision makers.

Center for Neighborhood Technology, http://www.cnt.org
>CNT invents and develops tools and methods for sustainable development.

Community Transportation Association of America, http://63.111.177.36
>CTAA is an association of organizations and individuals committed to improving mobility for all people.

Congress for the New Urbanism, http://www.cnu.org
>CNU is a collaboration of professionals working to reform North America's urban growth patterns.

Growth Management Institute, http://www.gmionline.org
>The Growth Management Institute is a small nonprofit organization established to encourage effective and equitable management of growth and change in human habitats. The Institute

promotes strategies and practices to achieve sustainable urban development and redevelopment in harmony with conservation of environmental qualities and features.

National Trust for Historic Preservation, http://www.nthp.org
> The National Trust provides information, technical assistance and advice to organizations and individuals working to preserve their communities and avoid urban sprawl. Concerned with the community disinvestment NTHP has conducted a wide variety of activities that include promotion of federal transportation and tax policies that encourage community revitalization and the rehabilitation of historic houses.

National Neighborhood Coalition, http://www.neighborhoodcoalition.org
> NNC promotes a neighborhood focus at all levels of government and throughout society by advocating for programs and policies that foster partnerships between neighborhood organizations, private sector institutions, and government agencies.

Natural Resources Defense Council, http://www.nrdc.org
> NRDC is a nonprofit organization with more than 400,000 members nationwide; its mission is to preserve the environment, protect the public health, and ensure the conservation of wilderness and natural resources.

The Northeast-Midwest Institute, http://www.nemw.org
> The Institute's Urban Environment Program explores the impacts of federal programs and policies at the local level, and educates key constituencies in its states and through the Northeast-Midwest Congressional and Senate Coalitions. Program staff have published a number of reports, case studies, and legislative summaries and matrices on the subjects of brownfields and smart growth.

Smart Growth Network, http://www.smartgrowth.org
> SGN encourages development that is environmentally, fiscally, and economically smart and helps create national, regional, and local coalitions to support smart growth.

Sustainable Communities Network, http://www.sustainable.org
> The mission of SCN is to connect individuals and organizations nationwide to the resources they need to help make their communities environmentally sound, socially equitable, and economically prosperous.

Tools for a Sustainable Community, http://www.iclei.org/la21/onestop.htm
> International Council for Local Environmental Initiatives (ICLEI) is the international environmental agency for local governments. Its mission is to build and serve a worldwide movement of local governments to achieve tangible improvements in global environmental and sustainable development conditions through cumulative local actions.

TransAct, http://www.transact.org
> The Transportation Action Network has provided The Surface Transportation Policy Project (STPP), the goal of which is to ensure that transportation policy and investments help conserve energy, protect environmental and aesthetic quality, strengthen the economy, promote social equity, and make communities more livable.

Transportation for Livable Communities, http://www.tlcnetwork.org
> TLCNet is a resource for people working to create more livable communities by improving transportation, and is intended to serve people working in city neighborhoods, suburbs, and rural

areas.

The Transportation Research Board, http://www.nas.edu/trb
   TRB is a unit of the National Research Council, a private, nonprofit institution that is the principal operating agency of the National Academy of Sciences and the National Academy of Engineering. Its mission is to promote innovation and progress in transportation by stimulating and conducting research, facilitating the dissemination of information, and encouraging the implementation of research results.

TRB Transportation and Air Quality Committee, http://transaq.ce.gatech.edu
   The Transportation Research Board Committee on Transportation and Air Quality has a website that provides information about the committee's activities, research, and membership. Its work includes examining the full range of relationships between transportation and air quality including regulatory and policy considerations, modeling practices, health effects, new technologies and transportation management strategies.

Trust for Public Land, http://www.tpl.org
   TPL helps agencies and communities protect land for parks, open space, and other public purposes by bringing real estate expertise, transaction skills, and innovative public financing tools to federal, state, local, and private partners.. TPL works in collaboration with communities and organizations around the country to bring private land into public ownership and to help create parks, greenways, riverways and to protect those traditional landscapes that define the character of where we live. TPL helps communities acquire endangered open space, create urban parks and promote bond issues to purchase open spaces.

Urban Land Institute, http://www.uli.org
   ULI is a nonprofit research and educational institute whose mission is to provide responsible leadership in the use of land in order to enhance the total environment.

Other sites of interest

Green Budget Reform Case Studies, http://iisd1.iisd.ca/greenbud/makingb.htm
   This website consists of twenty-three case studies from Europe and North America, providing lessons on measures that have been implemented to reduce environmental impacts and increase sustainability.

National Personal Transportation Survey, http://www-cta.ornl.gov/npts/1995/Doc/index.shtml
   The NPTS and the American Travel Survey (ATS) are household-based travel surveys conducted every five years by the DOT. The emphasis of the NPTS is on daily, local trips while the emphasis of the ATS is on long-distance travel in the United States.

Urbanized Area Formula Grants Program,
http://www.livablecommunities.gov/toolsandresources/tr_grants.htm
   Sponsored by FTA, this program provides funding for transit capital projects, such as buses, and assistance for operating expenses to urbanized areas with a population of 50,000 or more.

APPENDIX C RELATED WORK EFFORTS

Comparing Methodologies to Assess Transportation and Air Quality Impacts of Brownfields and Infill Development

This work, performed by ICF Consulting under contract with the EPA Office of Policy, describes four possible methodologies for including the emissions reductions produced by brownfields redevelopment and infill development in State Implementation Plans (SIPs). Each methodology examined here provides a different answer to the question: if the infill development for which emissions credit is being claimed had not been built, where would the development —the "growth increment"—have gone instead?

Methodology 1: Growth would have gone to a single "typical" greenfield site

Methodology 2: Growth would have gone to the fastest-growing parts of the region

Methodology 3: Growth would have been distributed throughout the region, in amounts determined by the local land use model

Methodology 4: Growth would have been distributed throughout the region, in amounts proportional to the distribution of all other growth.

EPA is also currently investigating similar methodologies for quantifying Transit Oriented Development under an ongoing contract with ICF.

This document is available at the following web address: http://www.epa.gov/oms/transp/traqsusd.htm under EPA report number EPA 231-R-01-001.

A Methodology to Establish SIP Creditability of Infill Development

This work was conducted by Apogee/Hagler Bailly and Criterion under EPA's Office of Policy (OP), which preceded the work by ICF mentioned above. Preliminary work performed is described in a draft report entitled *The Transportation and Environmental Impacts of Infill versus Greenfield Development: A Comparative Case Study Analysis.*

This study uses regional travel demand modeling to compare the travel and emissions impacts of a hypothetical development located on an infill site versus a greenfield site. Models were run for three case studies, in San Diego, California; Montgomery County, Maryland; and West Palm Beach, Florida.

Each case study consisted of modeling a hypothetical large development as if it were located on an actual infill site, and then modeling the same development as if it were on an actual greenfield site. The development size remains the same in both locations, but the density and street patterns are consistent with the surrounding urban form at each location. In each case, the MPO travel demand model was used to simulate the travel impacts of the development. Environmental impacts (including NOx and CO_2 emissions) and energy use were estimated using a GIS-based model called INDEX.

All three case studies show that locating the development on the infill site results in lower vehicle use and lower vehicle emissions. VMT per capita at the infill sites was roughly half that at the greenfield sites. NOx emissions were 27 percent to 42 percent lower at the infill sites, even though congestion at one infill site was higher than the greenfield site. It should be noted that the INDEX model uses simplified per-mile and per-trip emissions factors, not the standard vehicle emissions models.

Transportation Impacts of Micro-Scale Urban Design Elements: Data Collection and Modeling Needs

This 1998 joint DOT (FHWA)/EPA (OTAQ and OP) funded project brings together current knowledge and recent research concerning the ability to appropriately reflect the transportation impacts of various micro-scale urban design elements (e.g., sidewalk width, building setback, street grid type, etc.). The report explains procedures to estimate how land use development strategies and site design elements affect travel behavior and gives examples from selected MPO experience. Particularly useful for MPOs is a product that will relate specific urban design changes to auto ownership, trip generation (or tour or activity generation), and mode choice for use in current travel demand models. For copies contact the Federal Highway Administration and reference report number DTFH61-95-C-00168.

Air Quality Impacts of Regional Land Use Policies

Published in February 2000, this document, developed for the US EPA by Robert Johnson of the University of California, and John E. Abraham of the University of Calgary illustrates the air quality benefits or deficits of regional policy scenarios that affect land use development patterns. The objective of this study was to evaluate urban transportation scenarios in a mid-size region, (Sacramento, CA) that would significantly reduce vehicle emissions. A set of policies that include transit development, transit oriented design, and auto pricing were identified as particularly promising for the reduction of regional emissions over a 10- to 20-year time horizon. Key conclusions are that it may be important to model the land use effects of transportation scenarios; integrated land use and transportation models can provide important policy insights; land use intensification measures accompanied by supportive transit and/or pricing can produce comparatively large reductions in VMT and auto emissions; and that HOV and HOT lanes may increase vehicle emissions.

The Effects of Urban Form on Travel and Emissions: A Review and Synthesis of the Literature

This is an ongoing contract with Apogee/Hagler Bailly under EPA's Office of Policy (OP). The draft report offers a thorough summary of recent research on the effect of land use on travel behavior. Studies fall into two general categories. Empirical studies compare data collected from actual communities and try to distinguish how various land use factors lead to different travel patterns. Simulation studies use computer models to examine the impact of hypothetical land use patterns on travel and emissions.

The report concludes that changes in land use can reduce region-wide vehicle use and emissions over a period of several decades. Using simulation models, several studies have convincingly shown that modifying future development patterns in ways that make them less dependent on automobile use will reduce VMT and emissions. The reduction in emissions comes from shorter trip lengths and shifts to transit, bicycling, and walking modes. While computer modeling has improved greatly in recent years, it is still subject to some serious limitations. Zonal size generally precludes modeling the impact of micro-scale design features, for example.

The report documents how numerous empirical studies have shown relationships between specific land use factors and components of travel demand. For example, compact clusters of mixed-use development are correlated with reduced trip lengths. Similarly, higher density communities of mixed land use are associated with higher shares of travel by transit, bicycling and walking. The report acknowledges the methodological flaws that limit the conclusions that can be drawn from empirical studies. Some, for example, do not control for factors like income when comparing neighborhoods. A more fundamental flaw is the fact that cross-sectional studies, by nature, cannot establish causality.

Evaluation of Modeling Tools for Assessing Land Use Policies and Strategies

This complementary effort was done for the EPA Transportation and Market Incentives Group by Systems Application International (SAI). Its final report was issued in August 1997. The work was intended to assess how regional land use forecasting models are able to incorporate specific land use policies. The report evaluates three commercial land use models: DRAM/EMPAL, MEPLAN, and TRANUS. Each model was evaluated in terms of how well it could account for policies designed to (1) increase development densities, (2) increase land use mixing, and (3) modify design elements and infrastructure to encourage alternative travel modes. The specific policies used to achieve these goals were summarized as zoning, monetary incentives (such as subsidies to developers to build in targeted areas), and non-monetary incentives (such as reduced parking requirements).

The study concludes that DRAM/EMPAL, because it does not easily represent costs, cannot model the impact of any of the three types of policies. MEPLAN and TRANUS do include representations of development costs, and therefore can at least partially model zoning policies as well as monetary and non-monetary incentives. The report points out that all the models are seriously constrained by zonal size, however. They are usually run using zones the size of several census tracts, or a single census tract at the smallest. As a typical urban census tract is roughly one square mile, a model built on zones of this size could possibly detect an increase in density within a half-mile of a transit station or transit corridor; it could not detect smaller-scale land use changes. If the zonal system uses aggregations of census tracts, even transit station-area densities could not be resolved.

(EPA Report number EPA 420-R-97-007. Copies of the document can be obtained by calling 1-800-490-9198 and citing the EPA reference number.)

Green Development with the National Association of Home Builders Research Center

Past research has demonstrated that the location of new development has a strong correlation with its environmental impact. For example, far-flung residential subdivisions will generate more air pollution than "close-in" mixed use communities.

Beyond the location of development (far-flung versus close-in), research demonstrates that applying various development practices and techniques to the development site can mitigate environmental impacts. For example, proximity to transit, reduction in street width, grid lay-out for streets, construction of sidewalks and pathways, and co-location of diverse land uses (e.g., residential and commercial) will all improve community walkability and help reduce reliance on car use. The EPA Transportation and Market Incentives Group's cooperative agreement with the NAHB Research Center will generate a compendium of various development practices and techniques that create a range of environmental benefits. The "guidebook" will provide information on various green development techniques and will describe the steps for developing local, community-based green development programs. The development guide is being refined through a partnership with the Denver Home Builders Association, which is currently creating a "Green Developers' Program" for local builders.

NAHB can be contacted at http://www.nahb.com/ or National Association of Home Builders, 1201 15th street, NW Washington, DC 20005, phone: 800-368-5242 or 202-822-0200 in theWashington, D.C. area.

Projecting Land-Use Change: A Summary of Models for Assessing the Effects of Community Growth and Change on Land-Use Patterns

Many potential clients for land-use change models, such as city and county planners, community groups, and environmental agencies, need better information on the features, strengths, and limitations of various model packages. Because of this growing need, EPA has developed a selective inventory and evaluation of 25 leading land-use change models currently in use or under development. Partners in scoping this effort included the U.S. Department of Transportation, the Department of Interior, the academic and consulting communities, and multiple program offices across EPA.

EPA's Office of Research and Development (ORD) initiated the land-use change models evaluation in order to improve its ability to assess and mitigate future risk to ecological systems, human health, and quality of life. Land-use change is perhaps the most significant source of adverse impacts to aquatic and terrestrial environments today. Through its Regional Vulnerability Assessment and other initiatives, ORD is considering land-use change models at nested spatial scales in order to target ecological resources and socioeconomic issues for community-based protection efforts. The strategic evaluation of leading models achieves the following ORD objectives:

- Identify models that are immediately available for application at multi-county and watershed scales in rural and urban areas;

- Evaluate model frameworks for their ability to support alternative algorithms currently under development within ORD; and

- Assess the suite of inventoried models for gaps and weaknesses that ORD may seek to address through in-house research and external research grants.

Target user groups for the land-use change models evaluation are:

- Community planners and citizens who are seeking tools to analyze future land-use scenarios;

- EPA program office and regional staff who support communities with smart-growth planning tools and information; and

- ORD modelers and research planners who are currently assessing land-use models and gaps in the state of the science.

The document reference information is: U.S. EPA, 2000. Projecting Land-Use Change: A Summary of Models for Assessing the Effects of Community Growth and Change on Land-Use Patterns. EPA 600-R-00-098. U.S. EPA, Office of Research and Development, Cincinnati, OH. 260 pp. Copies of the document can be obtained by calling 1-800-490-9198 and citing the EPA reference number. For more information, please contact Laura Jackson at jackson.laura@epa.gov or (919) 541-3088.

APPENDIX D GLOSSARY OF TERMS

Accounting for air quality impacts of land use activities

The documented reductions in mobile source emissions due to land use activities that nonattainment and maintenance areas can use in their State Implementation Plans (SIPs) or conformity determinations.

Attainment area

An area considered to have air quality that meets or exceeds the U.S. Environmental Protection Agency's health standards used in the Clean Air Act. An area may be an attainment area for one pollutant and a non-attainment area for others.

Brownfields

Abandoned, idled or under-used industrial and commercial facilities where expansion or redevelopment is complicated by real or perceived environmental contamination.

Conformity

Process to assess the compliance of any transportation plan, program or project with air quality control plans. Conformity process is required by the Clean Air Act (CAA).

Control strategy

These are specific strategies for controlling the emissions, and reducing ambient levels, of pollutants in order to satisfy CAA requirements for demonstrations of reasonable further progress and attainment.

Criteria pollutants

Criteria pollutants are carbon monoxide (CO), lead (Pb), nitrogen dioxide (NO_2), ozone (O_3), particulate matter (PM), and sulfur dioxide (SO_2).

Economic Incentive Program (EIP)

Strategies that encourage emissions reductions through market based incentives and informational tools.

Infill development

A type of land use strategy. Specifically any type of new development that occurs within existing built-up areas (may be urban or suburban); includes brownfield development.

Initial forecast of future emissions

The level of emissions in the future that will result if no additional control measures are implemented other than what is required by law.

Jobs/housing balance

A type of land use strategy. Changes that reduce the disparity between the number of residences and the number of employment opportunities available within a sub-region.

Land use activity

Land use activities include all of the various actions that state and local governments or other entities take which affect the patterns of land use in a community or region. These activities result in patterns of land use that impact people's ability to travel. In this guidance, *land use activities* that reduce reliance on motor vehicles and increase accessibility of alternative modes of transportation and can be shown to have air quality benefits (e.g., through shortening trip lengths or increasing accessibility of alternative modes of transportation) can be accounted for in the air quality and transportation planning processes.

Land use activities include ***land use policies***, defined as specific policies, programs, or regulations adopted or administered by government agencies to allow and/or to encourage land uses that may result in decreased vehicle miles traveled and emissions of air pollutants and ***land use projects*** defined as specific developments that may be shown to reduce vehicle travel and emissions.

Maintenance area

Any geographic region of the United States previously designated nonattainment pursuant to the CAA Amendments of 1990 and subsequently redesignated to attainment subject to the requirement to develop a maintenance plan.

Mixed-use development

A type of land use strategy. Development that locates complementary land uses such as housing, retail, office, services and public facilities within walking distance of each other .

Mobile sources

A category of emission sources. This category includes motor vehicles, such as cars, light trucks, heavy duty trucks, and buses; locomotives; aircraft; construction equipment; lawn and garden equipment; boats and personal watercraft, etc.

National Ambient Air Quality Standards (NAAQS)

Standards for pollutants considered harmful to public health and the environment. NAAQS have been set for the six criteria pollutants (carbon monoxide, lead, nitrogen dioxide, ozone, particulate matter less than or equal to 10 microns in size, and sulfur dioxide).

Neotraditional development

A type of land use strategy. Specifically a set of land development and urban design elements intended to encourage and create pedestrian-oriented neighborhoods.

New Source Review

Air pollution permits are required for businesses that build new pollution sources or make significant changes to existing pollution sources. These are sometimes referred to as "preconstruction" or "new source review" permits. These permits are required to ensure that large new emissions do not cause significant health or environmental threats and that new pollution sources are well-controlled.

Nonattainment area

Any geographic region of the United States that has been designated as nonattainment for any pollutant for which a national ambient air quality standard exists.

Regionally significant

A term which has been defined in federal transportation planning regulations as applying to a transportation project that is on a facility that serves regional transportation needs (such as access to and from the area outside of the region, major activity centers in the region, major planned developments such as new retail malls, sports complexes, etc. or transportation terminals as well as most terminals themselves) and would normally be included in the modeling of a metropolitan area's transportation network, including at a minimum all principal arterial highways and all fixed guideway transit facilities that offer an alternative to regional highway travel.

State Implementation Plan (SIP)

State air quality plans required by the Clean Air Act for nonattainment and maintenance areas. The plans are prepared by state air quality agencies and include estimates of future air quality and control strategies to attain appropriate air quality standards.

Transit oriented development (TOD)

A type of land use strategy. This development encourages moderate to high density development along a regional transit system.

Transportation control measure (TCM)

Encompasses elements of both transportation system management (TSM) and transportation demand management (TDM). Transportation system management generally refers to the use of low capital intensive transportation improvements to increase the efficiency of transportation facilities and services. These can include carpool programs, parking management, traffic flow improvements, high occupancy vehicle lanes, and park and ride lots. TDM generally refers to policies, programs, and actions that are directed towards decreasing the use of single occupant vehicles. TDM also can include activities to encourage shifting or spreading peak travel periods. In practice, there is considerable overlap among these concepts and TCM, TSM and TDM are often used interchangeably.

Travel analysis zone (TAZ)

Level of geographic detail used in most transportation planning applications to summarize socioeconomic characteristics and travel data. TAZs vary in size depending on density and homogeneity of land uses, and are defined by local agencies.

Travel demand model

In the field of transportation there is a standard set of planning methods and models that are called the four-step process or the Urban Transportation Planning System (UTPS for short). This set of models and procedures is used to forecast travel demand for future transportation systems, and it plays a central role in the evaluation of alternative transportation plans and policies.

Vehicle miles traveled (VMT)

The number of miles driven in a certain area over a period of time. VMT can be reported as a per capita average, or as an aggregate number to reflect the total travel in an area over some time period.

Voluntary Mobile Source Emission Reduction Programs (VMEP)

Voluntary emission reduction programs that rely on the actions of individuals or other parties for achieving emissions reductions. The VMEP Policy is intended to provide an incentive for states, localities, and the public to voluntarily reduce air pollution in their communities.

Appendix E   List of Acronyms

CMAQ	Congestion Mitigation and Air Quality Improvement Program
CO	Carbon Monoxide
COG	Council of Governments
EIP	Economic Incentive Program
EPA	Environmental Protection Agency
MPO	Metropolitan Planning Organization
NAAQS	National Ambient Air Quality Standards
NO_2	Nitrogen Dioxide
NO_x	Oxides of Nitrogen
PM-10	Particulate Matter (10 micrometers or less)
SIP	State Implementation Plan
TCM	Transportation Control Measure
TOD	Transit-oriented development
VMEP	Voluntary Mobile Source Emissions Reduction Programs
VMT	Vehicle Miles Traveled

Appendix F   References to Relevant Policies, Guidance Documents, and General Information Sources

Granting Air quality credit Land Use Measures: Policy Options, 1999 (EPA 420-P-99-028)
http://www.epa.gov/oms/transp/traqsusd.htm

Background Information for Land Use SIP Policy, 1998 (EPA 420-R-98-012)
http://www.epa.gov/oms/transp/traqsusd.htm

Evaluation of Modeling Tools for Assessing Land Use Policies and Strategies, 1997 (EPA 420-R-97-007)
http://www.epa.gov/oms/transp/traqsusd.htm

Voluntary Emission Reduction Programs Guidance, 1997 (Memorandum)
http://www.epa.gov/oms/transp/vmweb/vmpoldoc.htm

Economic Incentive Program Guidance, 2001 (EPA 451/R-01-001)
http://www.epa.gov/ttn/oarpg/new.html
or
http://www.epa.gov/ttn/ecas/

Conformity Rule and supplemental documentation
http://www.epa.gov/oms/transp/traqconf.htm

Appendix G   Regional and State Contacts

Below is a listing of organizations that may be contacted in order to find out what agencies are responsible for the conformity and/or State Implementation Planning process in any given geographic area.

For State or Local Air Agencies

    State and Territorial Air Pollution Program Administrators/Association of
    Local Air Pollution Control Officials
    444 North Capitol St. N. W.
    Washington, D. C. 20001
    Telephone: 202-624-7864

For Metropolitan Planning Organizations or Councils of Government

    National Association of Regional Councils
    1700 K St. N. W.
    Washington, D. C. 20006
    Telephone: 202-457-0710

For Transit Agencies

    American Public Transportation Association
    1201 New York Avenue, N. W.
    Washington, D. C. 20005
    Telephone: 202-898-4000

For State Departments of Transportation

    American Association of State Highway and Transportation Officials
    444 N. Capitol St. N. W.
    Washington, D.C. 20001
    Telephone: 202-624-5800

For Environmental Protection Agency Contacts

    Questions on transportation conformity or a current listing of non-attainment and maintenance areas should be directed to:

    EPA Office of Transportation and Air Quality
    2000 Traverwood Drive
    Ann Arbor, MI 48105
    Telephone: 734-214-4441
    www.epa.gov/oms/transp/traqconf.htm

For EPA Regional Offices - Transportation Planning Contact *

Region I: Boston, MA
    617-918-1665 (RI, CT)
    617-918-1668 (MA, ME, VT, NH)

Region II: New York, NY
    212-637-3901 (NJ, Puerto Rico, U.S. Virgin Island)
    212-637-3804 (NY)

Region III: Philadelphia, PA
    215-814-2183 (DC, MD, VA)
    215-814-2184 (DE, PA, WV)

Region IV: Atlanta, GA
    404-562-9026 (AL, FL, GA, KY, MS, NC, SC, TN)

Region V: Chicago, IL
    312-353-8656 (Il, OH)
    312-353-4366 (IN)
    312-353-6680 (MI, MN, WI)

Region VI: Dallas, TX
    214-665-7247 (AR, LA, NM, OK, TX)

Region VII: Kansas City, KS
    913-551-7651 (IA, KS, MO, NE)

Region VIII: Denver, CO
    303-312-6446 (CO, MT, ND, SD, UT, WY)

Region IX: San Francisco, CA
    415-744-1247 (AZ, NV, CA)
    415-744-1231 (CA)
    415-744-1153 (CA)

Region X: Seattle, WA
    206-553-1463 (AK, ID, OR, WA)

***Please note:** This list is current as of publication date. For the most current list, visit EPA's web site at: www.epa.gov/epahome/locate2.htm or the TRAQ web site at: www.epa.gov/oms/transp/conform/contacts.htm.

www.ingramcontent.com/pod-product-compliance
Lightning Source LLC
Chambersburg PA
CBHW081729170526
45167CB00009B/3760